D1509169

Springer Series in Operations Research

Editors:
Peter W. Glynn   Stephen M. Robinson

Springer
New York
Berlin
Heidelberg
Barcelona
Hong Kong
London
Milan
Paris
Singapore
Tokyo

Springer Series in Operations Research

David D. Yao   Shaohui Zheng

Dynamic Control of Quality in Production-Inventory Systems

Coordination and Optimization

Springer

David D. Yao
Department of Operations Research and
   Industrial Engineering
Columbia University
New York, NY 10027
USA
yao@ieor.columbia.edu

Shaohui Zheng
Department of Information and
   Systems Management
Hong Kong University
   of Science and Technology
Clear Water Bay
Kowloon, Hong Kong
imzheng@ust.hk

Series Editors:
Peter W. Glynn
Department of Management Science
   and Engineering
Terman Engineering Center
Stanford University
Stanford, CA 94305-4026
USA
glynn@leland.stanford.edu

Stephen M. Robinson
Department of Industrial Engineering
University of Wisconsin–Madison
1513 University Avenue
Madison, WI 53706-1572
USA
smrobins@facstaff.wisc.edu

Mathematics Subject Classification (2000): 90C39, 62N10

ISBN 0-387-95491-0   Printed on acid-free paper.

Printed in the United States of America.

9 8 7 6 5 4 3 2 1   SPIN 10877263

Typesetting: Pages created by the authors using a Springer T$_E$X macro package.

www.springer-ny.com

Springer-Verlag   New York Berlin Heidelberg
A member of BertelsmannSpringer Science+Business Media GmbH

Acknowledgments

This book is based on our scientific collaborations that started in 1993/94, when Shaohui Zheng was completing his doctoral research at Columbia University. Subsequent collaborations took place as we exchanged visits between New York (Columbia University) and Hong Kong (University of Science and Technology, and the Chinese University). Over these years, our research has been supported by NSF grants DMI-952309, ECS-9705392, and DMI-0085124; RGC grant CUHK4376/99E; RGC/NSFC grant CUHK10 (DDY); and by RGC grants HKUST6220/97H, HKUST6012/00E, and HKUST6089/00E (SZ). We thank our home institutions and the funding agencies in the United States and Hong Kong for their sponsorship.

Chapters 3, 9, and 10 are based on joint research with Dr. Jinfa Chen (of Wall Street Systems, New York), an inspiring colleague and loyal friend, who has shared with us the travail and transcendence of discovery.

DDY & SZ

Contents

1
Introduction

Quality control, or more specifically, statistical process control (SPC), is about scientific means for conducting observations, tests and inspections and thereby making decisions that improve the performance of industrial processes. Deming [29] said it well:

> ... the only reason to carry out a test is to improve a process, to improve the quality and quantity of the next run or of next year's crop. Important questions in science and industry are how and under what conditions observations may contribute to a rational decision to change or not to change a process to accomplish improvements.

The key issue, therefore, is how and under what conditions monitoring and testing the process will lead to quality improvement.

To address this issue, we have in recent years developed a set of *dynamic* approaches, in the sense that they are concerned with finding *policies* or *controls*, instead of merely parametric or statistical designs (such as determining the upper and lower limits in control charts). We have focused our studies on identifying certain optimal sequential procedures and revealing their structural properties. For example, in several applications, we characterize the quality of the underlying production process by a random defective rate, Θ, known only by its distribution. As the process is monitored, the posterior probability of Θ is updated, and decisions are made accordingly—usually characterized by a sequence of thresholds. The obvious advantage of sequential procedures is that they make full and effective use of all the information resulting from sampling and testing.

1.1   Theme and Methodologies

A theme that characterizes our studies is *coordination*. In practice, quality control problems almost never exist in isolation. For example, in a multi-stage system, quality control at the upstream and downstream stages are highly coupled: passing defective items downstream is often more costly than correcting them upstream. A concerted effort is also needed between achieving larger quantity and better quality, so that the two will enhance, instead of interfere with, each other.

The basic methodology underlying our studies is the Markov decision process (MDP). We also incorporate into MDP the usage of stochastic comparison techniques, including those based on notions of stochastic convexity (Shaked and Shanthikumar [80]; Shanthikumar and Yao [84]) and stochastic submodularity (Chang, Shanthikumar and Yao [14]). For example, a key property underlying some of our main results is a strengthening of the usual submodularity (Topkis [98]) to what we call K-submodularity and its adaptation to the stochastic context. These properties play a crucial role in revealing the threshold structures of the policies and in proving their optimality.

The focal question that drives our studies is this: under what conditions–as general as possible, and for what systems–as broad as possible, does a certain class of policies become optimal, in the sense of striking the best coordination among several competing or even conflicting aspects of the logistics of batch production. We put particular emphasis on the class of policies that have simple threshold structures: they are simple enough to facilitate implementation, but sophisticated enough to be optimal.

As a result, in comparison to existing techniques (e.g., acceptance sampling), our approaches lead to optimal policies that explicitly and systematically take into account coordination. On the other hand, the threshold form of the optimal policies does resemble the structure of many existing techniques, and it shares the advantage of their readiness for implementation. For example, some of the optimal policies that we derived can be easily implemented as control charts by translating the sequence of thresholds into control limits.

1.2   Relations to Other Approaches

Here we present a brief review of the related literature and comment on the relationship between our studies and the general taxonomy of statistical quality control research.

MDP and sequential analysis in particular are standard and widely used techniques in stochastic control (e.g., Bertsekas [6]; Blackwell [8]; Chow, Robbins, and Siegmund [24]; Puterman [71]; Ross [74]; and Wald [100]. Specific applications of sequential analysis in SPC include sequential likelihood ratio tests and related sequential sampling and control chart techniques (see, e.g., Banks [4]; Coleman [25]; Mitra [60]; Montgomery [61]; Shewhart [86]; and Thompson and Koronacki [97]). Applications of stochastic control techniques to SPC are relatively few; some examples include Kalman filtering and multivariate techniques (e.g., Crowder [28]; Hubele [45]; Takata et al [93]). There are also a few studies that apply MDP to study the detection of state shift of machining processes and related issues such as replacement, process revision, and allocation of inspection effort (e.g., Girshick and Rubin [37]; Ross [75]; Taylor [95]; and White [102, 103]). Some of these earlier studies are sources of motivation for our work.

Our approaches put special emphasis on the relationship between the process control and other logistics aspects of batch production. (This relationship is, in fact, one of the elements in the QFD (quality function deployment) matrix of Deming [30].) On the other hand, our approaches do not easily fall into the dichotomy (e.g., [60]) of process quality control (via monitoring, testing, and revision) versus product quality control (via acceptance sampling techniques). Rather, we aim to unify the two: intelligently monitor the quality of the process to improve the quality of the process and hence the quality of the product, in the context of coordinating and balancing other aspects of production logistics. For example, under our sequential control policies usually the effort spent on inspection is controlled to the level necessary for process monitoring and quality assurance, and the saved time and effort (in terms of capacity) is used for more productive work, including process improvement.

In terms of process control, our approaches do not necessarily fit into the dichotomy of online versus off-line control (e.g., Mitra [60]; Taguchi, Elsayed, and Hsiang [91]; Taguchi and Wu [92]). To the extent that our approaches make extensive use of the real-time process dynamics, they appear to belong to the online category. However, to the extent the approaches exploit the interplay between process control and other production logistics, the optimal policies they generate clearly have direct implications on off-line designs as well. Indeed, some of the off-line parameters can be easily incorporated into the objective functions of our MDP formulation.

Throughout our studies we have assumed an indirect monitoring mechanism on the process, via inspection of the items produced. This is consistent with the status of today's technology: it remains extremely difficult, if not impossible, to do direct process sensing and monitoring (see more discussions along this line in Chapter 4). On the other hand, our model is readily adapted to the case of direct sensing, wherever the technology applies. For example, the threshold structure of a process control policy easily translates into a sensor-based implementation: revise the process whenever the process 'state', in terms of certain key parameters, is detected as exceeding the threshold values.

1.3   Organization and Overview

As mentioned earlier, the theme of our studies has been to address the coordination of quality control with other aspects of a firm's production system and supply chain. Specifically, we shall focus on the following issues:

- coordination between the inspection and repair of finished products and their follow-up services (Chapter 3);

- coordination between inspection and process revision (Chapter 4);

- interstage coordination:

  - between production and inspection (Chapter 5),
  - under capacity constraints (Chapter 6),
  - of the inspection processes at different stages (Chapter 7);

- coordination between the inspection of different components in an assembly system (Chapter 8);

- coordination between replenishment and rework decisions (Chapter 9).

Therefore, we start from the customer end, which deals with warranty or service contracts in Chapter 3, and move to the production/process control in Chapter 4. These are followed by coordination in *serial* stages of production in Chapters 5, 6, and 7, addressing different issues as highlighted earlier. In Chapter 8, we turn to a *parallel* configuration, an assembly system, and study the coordination of quality control among its components. This brings us to the other end of the supply chain in Chapter 9, where we study the coordination between replenishment (order quantities) and rework (quality improvement) if our suppliers are unreliable.

The subject of the last chapter, Chapter 10, is substitutable inventory systems. It relates to quality control in that both provide means of recourse after production is completed. Inspection is usually carried out *before* demand is realized, to improve, for example, the quality of finished products and reduce after-sales service costs. A substitutable inventory, on the other hand, allows recourse actions *after* demand is realized: the surplus of higher-end products can be used if necessary to substitute for the shortage of lesser products, to reduce any penalty cost associated with unmet demand or lost sales.

2
Stochastic Monotonicity, Convexity and Submodularity

In this chapter we collect preliminary material on various notions of stochastic comparisons and related techniques, most of which will be used in later chapters.

We start with stochastic ordering and likelihood ratio ordering in §2.1, focusing on their functional representations. In §2.2 we present several versions of stochastic convexity. This is followed by §2.3, where we discuss submodularity and supermodularity, their isotone properties, and their stochastic counterparts. Some applications in the context of Markov chains, birth and death processes in particular, are illustrated in §2.4.

Throughout this chapter and the rest of the book, the terms, increasing and decreasing, are used in the nonstrict sense to mean "nondecreasing" and "nonincreasing", respectively.

2.1 Stochastic and Likelihood-Ratio Orderings

Definition 2.1 X and Y are two random variables, with distribution functions $F(\cdot)$ and $G(\cdot)$.

(i) X and Y are ordered under *stochastic ordering*, denoted $X \geq_{st} Y$, if

$$\bar{F}(a) := 1 - \mathsf{P}[X \leq a] \geq 1 - \mathsf{P}[Y \leq a] := \bar{G}(a), \text{ for all } a.$$

(ii) X and Y are ordered under *likelihood ratio ordering*, denoted $X \geq_{lr} Y$, if

$$f(x)g(y) \geq f(y)g(x), \text{ for all } x \geq y,$$

where $f(\cdot)$ and $g(\cdot)$ are, respectively, the densities of X and Y, if they exist; or, in the case of discrete random variables, if

$$P[X = k]P[Y = k - 1] \geq P[X = k - 1]P[Y = k], \quad \text{for all } k.$$

Lemma 2.2 For two random variables X and Y, $X \geq_{st} Y$ if and only if $\mathsf{E}\phi(X) \geq \mathsf{E}\phi(Y)$ for all increasing functions ϕ.

Proof. Let $U \in [0, 1]$ denote the uniform random variable. Then

$$X \stackrel{\mathrm{d}}{=} F^{-1}(U), \qquad Y \stackrel{\mathrm{d}}{=} G^{-1}(U),$$

where $\stackrel{\mathrm{d}}{=}$ denotes equal in distribution, and F^{-1} and G^{-1} denote the inverse functions (of F and G). From Definition 2.1 (i), we know $X \geq_{st} Y$ means $F(a) \leq G(a)$ for any a or, equivalently,

$$F^{-1}(u) \geq G^{-1}(u), \quad \forall u \in [0, 1].$$

Therefore, for any increasing function ϕ, we have

$$\mathsf{E}\phi(F^{-1}(U)) \geq \mathsf{E}\phi(G^{-1}(U)),$$

and hence, $\mathsf{E}\phi(X) \geq \mathsf{E}\phi(Y)$. For the converse, simply pick ϕ to be the indicator function $\phi(x) := \mathbf{1}[x \geq a]$. Then $\mathsf{E}\phi(X) \geq \mathsf{E}\phi(Y)$ means $F(a) \leq G(a)$. $\square$

Lemma 2.3 For two random variables X and Y, $X \geq_{lr} Y \Rightarrow X \geq_{st} Y$.

Proof. We want to show $\bar{F}(a) \geq \bar{G}(a)$ for any a. Notice the following identities:

$$\bar{F}(a) = \exp\left[-\int_{-\infty}^{a} \frac{f(s)}{\bar{F}(s)} ds \right], \quad \bar{G}(a) = \exp\left[-\int_{-\infty}^{a} \frac{g(s)}{\bar{G}(s)} ds \right].$$

Hence, it suffices to show

$$f(s)\bar{G}(s) \leq \bar{F}(s)g(s).$$

But this follows from integrating on both sides of the following (with respect to t, over $t \geq s$):

$$f(s)g(t) \leq f(t)g(s), \qquad s \leq t,$$

which, in turn, follows from $X \geq_{lr} Y$. $\square$

For the likelihood ratio ordering to have a functional representation similar to that of stochastic ordering in Lemma 2.2, we need a *bivariate* function.

Proposition 2.4 Define a class of bivariate functions as follows:

$$\mathbf{C}_{lr} = \{\phi(x, y) : \Delta\phi(x, y) = \phi(x, y) - \phi(y, x) \geq 0, \forall x \geq y\}.$$

Then, for two *independent* random variables X and Y, $X \geq_{lr} Y$ if and only if $\mathsf{E}\phi(X, Y) \geq \mathsf{E}\phi(Y, X)$, for all $\phi \in \mathbf{C}_{lr}$.

Proof. For the "if" part, given $u > v$, let

$$\phi(x, y) = \mathbf{1}[u \leq x \leq u + du, v \leq y \leq v + dv],$$

where $du > 0$, $dv > 0$, and $dv < u - v$; then $\phi \in \mathbf{C}_{lr}$. Hence, $\mathsf{E}\phi(X,Y) \geq \mathsf{E}\phi(Y, X)$; or $f(u)g(v) \geq f(v)g(u)$, for $u > v$; hence, $X \geq_{lr} Y$.

For the "only if" part, taking into account $\Delta\phi(x, y) = -\Delta\phi(y, x)$, we have

$$
\begin{aligned}
\mathsf{E}\Delta\phi(X,Y) &= \int_y \int_x \Delta\phi(x, y) f(x)g(y) dx dy \\
&= \int_y \int_{x \geq y} \Delta\phi(x, y)[f(x)g(y) - f(y)g(x)] dx dy \\
&\geq 0,
\end{aligned}
$$

because

$$\phi \in \mathbf{C}_{lr} \;\Rightarrow\; \Delta\phi \geq 0$$

and

$$X \geq_{lr} Y \;\Rightarrow\; f(x)g(y) \geq f(y)g(x), \qquad \forall\, x \geq y. \qquad \square$$

Note that the likelihood ratio ordering (and the stochastic ordering) is a relation concerning the *marginal* distributions of X and Y. In other words, there is no reference to any joint distributional relationship of (X, Y). Hence, it is only natural to assume independence among the two random variables X and Y in the preceding theorem, which can be rephrased as follows.

Corollary 2.5 $X \geq_{lr} Y$ if and only if there exist two independent random variables $\hat{X}$ and $\hat{Y}$, such that $\hat{X} \stackrel{\mathrm{d}}{=} X$, $\hat{Y} \stackrel{\mathrm{d}}{=} Y$, and $\mathsf{E}\phi(\hat{X}, \hat{Y}) \geq \mathsf{E}\phi(\hat{Y}, \hat{X})$ for all $\phi \in \mathbf{C}_{lr}$.

Direct verification shows that the class $\mathbf{C}_{lr}$ is closed under composition with increasing functions. That is,

$$\phi \in \mathbf{C}_{lr} \;\Rightarrow\; h(\phi(x, y)) \in \mathbf{C}_{lr},$$

for any increasing function $h(\cdot)$.

Corollary 2.6 For two independent random variables, X and Y, $X \geq_{lr} Y$ if and only if $\phi(X, Y) \geq_{st} \phi(Y, X)$ for all functions $\phi \in \mathbf{C}_{lr}$.

2.2   Stochastic Convexity

The stochastic ordering in Definition 2.1 can be extended to a family of random variables parameterized by a real or integer-valued scalar θ, $X(\theta)$. We say $X(\theta)$ is stochastically increasing in θ, if $X(\theta_1) \leq_{st} X(\theta_2)$ for any $\theta_1 \leq \theta_2$. This is what we mean by *stochastic monotonicity*.

Definition 2.7 $X(\theta)$ is *stochastically increasing* in θ, if for any given $\theta_1 \leq \theta_2$, there exist on a common probability space $(\Omega, \mathcal{F}, P)$ two random variables $\hat{X}_1$ and $\hat{X}_2$ that are equal in distribution to $X(\theta_1)$ and $X(\theta_2)$, respectively, and $\hat{X}_1(\omega) \leq \hat{X}_2(\omega)$ for all $\omega \in \Omega$.

Similar to the proof of Lemma 2.2, suppose $F(x, \theta)$ denotes the distribution function of $X(\theta)$. Let $F^{-1}(u, \theta) = \inf\{x : F(x, \theta) > u\}$, with $u \in [0, 1]$, denote the inverse distribution function. Then the stochastic monotonicity defined earlier is equivalent to F^{-1} increasing in θ. The common probability space in this case is the one on which the uniform variate $U \in [0, 1]$ is defined. In particular, note that $F^{-1}(U, \theta_i) \overset{\mathrm{d}}{=} X(\theta_i)$ for $i = 1, 2$.

In view of Lemma 2.2, the following is immediate.

Definition 2.8 $X(\theta)$ is stochastically increasing in θ, if for any increasing function $\phi(\cdot)$, $\mathsf{E}\phi[X(\theta)]$ (as a *deterministic* function) is increasing in θ.

The preceding can be viewed as a "functional" definition for stochastic monotonicity, as opposed to the "sample-path" definition in 2.7. The two definitions are equivalent, in view of Lemma 2.2. The sample-path definition is easier to use in proofs (i.e., to establish stochastic monotonicity), and the functional version is more convenient in applications (for example, when ϕ is part of an objective function that is to be optimized).

Now consider the stochastic *convexity* of $\{X(\theta)\}$. Parallel to the two definitions for stochastic monotonicity, we have the following two definitions for stochastic convexity, which, however, are *not* equivalent, as we shall explain shortly.

Definition 2.9 (SICX-sp) $\{X(\theta)\}$ is stochastically increasing and convex in the sample-path sense, denoted $\{X(\theta)\} \in$ SICX-sp, if for any four parameter values θ_i, $i = 1, 2, 3, 4$, that satisfy $\theta_1 + \theta_4 = \theta_2 + \theta_3$ and $\theta_4 \geq \max\{\theta_2, \theta_3\}$, there exist on a common probability space $(\Omega, \mathcal{F}, P)$ four random variables $\hat{X}_i$, $i = 1, 2, 3, 4$, such that $\hat{X}_i \overset{\mathrm{d}}{=} X(\theta_i)$, for all four i, and

$$\hat{X}_4(\omega) \geq \max\{\hat{X}_2(\omega), \hat{X}_3(\omega)\}, \quad \hat{X}_1(\omega) + \hat{X}_4(\omega) \geq \hat{X}_2(\omega) + \hat{X}_3(\omega)$$

for all $\omega \in \Omega$.

Definition 2.10 (SSICX) $\{X(\theta)\}$ is said to satisfy *strong* stochastic increasing convexity, denoted $\{X(\theta)\} \in$ SSICX, if $X(\theta)$ can be expressed as $X(\theta) \overset{\mathrm{d}}{=} \phi(\xi, \theta)$, where ϕ is an increasing and convex function with respect to θ and ξ is a random variable whose distribution function is independent of θ. Specifically, $\{X(\theta)\} \in$ SSICX if $F^{-1}(u, \theta)$ is increasing and convex in θ (for any given $u \in [0, 1]$).

For completeness, we present a third version of stochastic convexity.

Definition 2.11 (SICX) $\{X(\theta)\}$ is stochastically increasing and convex in θ, denoted $\{X(\theta)\} \in$ SICX, if $E\phi[X(\theta)]$ is increasing and convex in θ for any increasing and convex function $\phi(\cdot)$.

The following implication relations are readily verified from the definitions.

Proposition 2.12 SSICX $\Rightarrow$ SICX-sp $\Rightarrow$ SICX.

Definitions for stochastic increasing *concavity* (SICV) are similarly defined. Also, stochastic *linearity* is defined as when both stochastic convexity and concavity hold. We shall use CV and L (replacing CX in the acronyms) to denote concavity and linearity, respectively. Also note that it is *not* necessary to define the stochastic increasingness together with stochastic convexity. The increasingness in the definitions can be taken out or replaced by decreasingness.

Remark 2.13 (i) The strong version of stochastic increasing convexity, SSICX, can also restated as follows: for any parameter values θ and η and any constant $\alpha \in [0, 1]$, there exist on a common probability space $(\Omega, \mathcal{F}, P)$ three random variables,

$$\hat{X}_1 \stackrel{d}{=} X(\theta), \quad \hat{X}_2 \stackrel{d}{=} X(\eta), \quad \hat{X}_3 \stackrel{d}{=} X(\alpha\theta + (1-\alpha)\eta),$$

such that

$$\hat{X}_3(\omega) \le \alpha\hat{X}_1(\omega) + (1-\alpha)\hat{X}_2(\omega)$$

for all $\omega \in \Omega$. In this sense, the sample values of $X(\omega, \theta)$ behave like a convex function in θ. This version mimics the convexity definition of a deterministic function. In contrast, SICX-sp relaxes the regular convexity definition to a four-point version and hence a weaker version of stochastic convexity.

(ii) SSICX is satisfied by many families of random variables. For example, if $X(\theta)$ is an exponential random variable with mean θ, then $X(\theta) = \theta\xi$ where ξ follows the exponential distribution with unit mean. Hence, $\{X(\theta)\} \in$ SSIL. In fact, this property extends to summations and mixtures of exponentials. Similarly, if $X(\theta)$ is a normal variate with mean μ and standard deviation θ, then $X(\theta) = \theta Z + \mu$, where Z denotes the standard normal variate. Hence, in this case we also have $\{X(\theta)\} \in$ SSIL, for fixed μ.

(iii) When the distribution function $F(x, \theta)$ is differentiable, it is also easy to verify SSICX, following Definition 2.10, by checking the first and second derivatives of X with respect to θ. To this end, write X as $X(\xi, \theta)$, and for each sample value of ξ, view X as a deterministic function of θ. Omitting reference to ξ and taking derivatives from $F(X(\theta), \theta) = U$, we have

$$F_x X_\theta + F_\theta = 0, \quad F_x X_{\theta\theta} + F_{xx} X_\theta^2 + 2F_{x\theta} X_\theta + F_{\theta\theta} = 0,$$

where F_x, F_θ, X_θ, etc. denote partial derivatives with respect to x or θ. From the preceding, solve for X_θ and $X_{\theta\theta}$. Then $\{X(\theta)\} \in$ SSICX if both derivatives are nonnegative.

The following examples illustrate that certain useful families of random variables do *not* satisfy SSICX, although they do satisfy the SICX-sp property.

Example 2.14 Let $X(p)$ denote a Bernoulli random variable: $X(p) = \mathbf{1}[U \leq p]$, where $p \in (0, 1)$ is the parameter, and $U \in [0, 1]$ is the uniform variate. Because $\mathsf{E}\phi[X(p)] = \phi(1)p$ is a linear function of p (regardless of the properties of ϕ), $X(p)$ satisfies stochastic linearity in terms of the weakest definition in 2.11. It is also clear that $X(p)$ does not satisfy SSICX (or SSICV), because the indicator function is neither convex nor concave.

The nonobvious fact here is that $X(p)$ satisfies SIL-sp in the sense of Definition 2.9. To show this, consider four parameter values: $p_1 \leq p_2 = p_3 \leq p_4$ and $p_1 + p_4 = p_2 + p_3$. Let

$$\hat{X}_i = \mathbf{1}[U \leq p_i], \quad i = 1, 2, 4,$$

and

$$\hat{X}_3 = \mathbf{1}[U \leq p_1] + \mathbf{1}[p_2 \leq U \leq p_4].$$

Then, clearly $\hat{X}_i \stackrel{\mathrm{d}}{=} X(p_i)$ for all four i. In particular $\hat{X}_3 \stackrel{\mathrm{d}}{=} \mathbf{1}[U \leq p_3]$, because $p_4 - p_2 + p_1 = p_3$. Also, $\hat{X}_4 \geq \max\{\hat{X}_2, \hat{X}_3\}$ and

$$
\begin{aligned}
\hat{X}_1 + \hat{X}_4 &= \mathbf{1}[U \leq p_1] + \mathbf{1}[U \leq p_2] + \mathbf{1}[p_2 \leq U \leq p_4] \\
&= \hat{X}_2 + \hat{X}_3.
\end{aligned}
$$

Hence, $\{X(p)\} \in$ SIL-sp.

Example 2.15 A similar example is when $X(p)$ follows a geometric distribution with parameter p. Because $\mathsf{E}X(p) = 1/p$, $\{X(p)\}$ is stochastically decreasing and convex in the weakest sense of Definition 2.11. We can further show that $\{X(p)\}$ is stochastically decreasing and convex in the sample-path sense of Definition 2.9. Consider the four p_i values as in the last example. Use a common sequence of i.i.d. uniform variates $\{U_n \in [0, 1]; n = 1, 2, ...\}$ to generate the four replicas:

$$
\begin{aligned}
\hat{X}_i &= \inf\{n : U_n \in [0, p_i]\}, \quad i = 1, 2, 4, \\
\hat{X}_3 &= \inf\{n : U_n \in [0, p_1] \cup [p_2, p_4]\}.
\end{aligned}
$$

Then we have $\hat{X}_1 \geq \hat{X}_i$ for $i \neq 1$, and $\hat{X}_4$ is equal to either $\hat{X}_2$ or $\hat{X}_3$; hence, $\hat{X}_1 + \hat{X}_4 \geq \hat{X}_2 + \hat{X}_3$.

Example 2.16 Let $X(n) = \sum_{i=1}^{n} Y_i$, where Y_i are i.i.d. nonnegative random variables. We show $\{X(n)\} \in$ SIL-sp. To this end, pick four parameter values: $n, n+1, n+1$, and $n+2$. Use a sequence of i.i.d. uniform variates, $\{U_1, ..., U_n, U_{n+1}, U_{n+2}\}$ to generate, from the common distribution of Y_i, the sample values $\{\hat{Y}_1, ..., \hat{Y}_n, \hat{Y}_{n+1}, \hat{Y}_{n+2}\}$. Let

$$\hat{X}_1 = \sum_{i=1}^{n} \hat{Y}_i, \quad \hat{X}_2 = \sum_{i=1}^{n+1} \hat{Y}_i, \quad \hat{X}_4 = \sum_{i=1}^{n+2} \hat{Y}_i,$$

and

$$\hat{X}_3 = \sum_{i=1}^{n} \hat{Y}_i + \hat{Y}_{n+2}.$$

The required conditions in Definition 2.9 are obviously satisfied by this construction. In particular, $\hat{X}_3 \stackrel{d}{=} X(n+1)$, because the $\hat{Y}_i$s are i.i.d.; and $\hat{X}_1 + \hat{X}_4 = \hat{X}_2 + \hat{X}_3$. It is also clear that $X(n)$ is neither convex nor concave in n, in the strong sense of SSICX.

Example 2.17 Continue with the preceding example. Suppose Y_i are independent, but not identically distributed. Specifically, suppose $Y_i \leq_{st} Y_{i+1}$ for all i. Then, $\{X(n)\} \in$ SICX-sp. To show this, modify the construction of $\hat{X}_3$ to the following:

$$\hat{X}_3 = \sum_{i=1}^{n} \hat{Y}_i + \hat{Y}'_{n+1},$$

where $\hat{Y}'_{n+1}$ is sampled from the distribution of Y_{n+1}, but the uniform variate used is U_{n+2}, which generates Y_{n+2}. Because $Y_{n+1} \leq_{st} Y_{n+2}$, we have $\hat{Y}'_{n+1} \leq \hat{Y}_{n+2}$. Hence, $\hat{X}_1 + \hat{X}_4 \geq \hat{X}_2 + \hat{X}_3$. Similarly, if $Y_i \geq_{st} Y_{i+1}$ for all i, then $\{X(n)\} \in$ SICV-sp.

Combining the two examples in 2.14 and 2.16, we have what is usually called a random yield model. Let $Y_i = Y_i(p)$, $i = 1, ..., n$, be a set of i.i.d. Bernoulli random variables parameterized by p. Suppose p is the yield rate or, $1 - p$ the defective rate. Then Y_i represents whether the ith unit, in a batch production of size n, is good ($Y_i = 1$) or bad ($Y_i = 0$); and X gives the total number of good units (i.e., "yield") that come out of this batch.

2.3   Stochastic Submodularity

Recall that, a bivariate function $\phi(x, y)$ is called *submodular* if for any $x_1 \leq x_2$, $y_1 \leq y_2$,

$$\phi(x_1, y_1) + \phi(x_2, y_2) \leq \phi(x_1, y_2) + \phi(x_2, y_1).$$

The function ϕ is called *supermodular* if $-\phi$ is submodular. Note that the preceding definition extends in the natural way to multivariate functions, for which all the following results also hold.

When the function ϕ is twice differentiable, then, following the preceding definition, submodularity is equivalent to the cross derivative $\frac{\partial^2 \phi}{\partial x \partial y} \leq 0$.

The minimizer of a submodular function has an "isotone" property as follows. Suppose that given y, we have

$$x^*(y) = \arg\min_x \phi(x, y).$$

Then, $x^*(y)$ is increasing in y. To see this, consider $y_1 < y_2$ and let $x_i^* := x^*(y_i)$ for $i = 1, 2$. Then, for any $x_2 < x_1^*$, submodularity implies that

$$\phi(x_2, y_2) - \phi(x_1^*, y_2) \geq \phi(x_2, y_1) - \phi(x_1^*, y_1).$$

Because the right-hand side is ≥ 0, due to the minimality of x_1^*, we have $\phi(x_2, y_2) \geq \phi(x_1^*, y_2)$. From this we can conclude that $x^*(y_2) \geq x_1^*$. Similarly, the maximizer of a supermodular function also has this isotone property; whereas the maximizer of a submodular function has the "antitone" (decreasing) property, and so does the minimizer of a supermodular function.

Definition 2.18 $X(s, t)$, where s and t are real or integer-valued scalars, is *stochastically supermodular* (resp. *submodular*) in (s, t) if for any given $s_1 \leq s_2$, $t_1 \leq t_2$, there exist on a common probability space $(\Omega, \mathcal{F}, P)$ four random variables $\hat{X}_i$, $i = 1, 2, 3, 4$, that are equal in distribution to $X(s_1, t_1), X(s_1, t_2), X(s_2, t_1)$, and $X(s_2, t_2)$, respectively, and

$$\hat{X}_1(\omega) + \hat{X}_4(\omega) \geq (\text{resp. } \leq)\hat{X}_2(\omega) + \hat{X}_3(\omega).$$

In light of its deterministic counterpart, the preceding definition simply states that the sample realization of X (i.e., for any given ω) behaves as a deterministic supermodular or submodular function.

Example 2.19 Continue with the yield model introduced at the end of the last section. It is readily verified that $X(n, p) = \sum_{i=1}^n Y_i(p)$, where $Y_i(p)$ are i.i.d. Bernoulli random variables with parameter p, is stochastically supermodular in (n, p). In fact, the Bernoulli distribution here is not important; supermodularity holds as long as for each i, $Y_i(p)$ is stochastically increasing in p. This supermodularity, as well as the SIL-sp property in n and in p established earlier, will prove to be very useful later when we study an assembly operation with random component yield.

Next, we combine (stochastic) supermodularity and convexity into what is called *directional convexity*.

Definition 2.20 A (deterministic) bivariate function $\phi(x, y)$ is called *directionally convex* (resp. *directionally concave*) if it satisfies the following properties:

$$\phi(z^1) + \phi(z^4) \geq (\text{resp. } \leq) \phi(z^2) + \phi(z^3),$$

for any $z^i = (x_i, y_i)$, $i = 1, 2, 3, 4$, such that

$$x_1 \leq x_2, x_3 \leq x_4, \quad y_1 \leq y_2, y_3 \leq y_4$$

and

$$x_1 + x_4 = x_2 + x_3, \quad y_1 + y_4 = y_2 + y_3.$$

Lemma 2.21 The bivariate function $\phi(x, y)$ is directionally convex (resp. directionally concave) in (x, y) if and only if ϕ is convex (resp. concave) in x and in y and supermodular (resp. submodular) in (x, y).

Proof. Consider the case of directional convexity, the case of directional concavity is completely analogous. Note that the four points z^i $(i = 1, 2, 3, 4)$ in Definition 2.20 have the following relation: z^1 and z^4 are at the southwest and northeast corners of a rectangle, while z^2 and z^3 lie within the rectangle. Hence, directional convexity implies supermodularity by pushing z^2 and z^3 to coincide with the other two corner points; and convexity (in each variable) follows by collapsing the rectangle to one of its sides, parallel to either the horizontal or vertical axis.

For the converse, consider the four points z^i $(i = 1, 2, 3, 4)$ in Definition 2.20. We have

$$
\begin{aligned}
& \phi(x_4, y_4) - \phi(x_3, y_3) \\
= \; & \phi(x_4, y_4) - \phi(x_4, y_3) + \phi(x_4, y_3) - \phi(x_3, y_3) \\
\geq \; & \phi(x_4, y_2) - \phi(x_4, y_1) + \phi(x_4, y_1) - \phi(x_3, y_1) \\
\geq \; & \phi(x_2, y_2) - \phi(x_2, y_1) + \phi(x_2, y_1) - \phi(x_1, y_1) \\
= \; & \phi(x_2, y_2) - \phi(x_1, y_1),
\end{aligned}
$$

where the first inequality makes use of convexity in y and supermodularity; and the second inequality makes use of convexity in x and supermodularity. $\square$

In view of this above lemma, we can combine the stochastic supermodularity with the various versions of stochastic convexity defined in the last section, to come up with the definition of stochastic directional convexity. The following is the sample-path version.

Definition 2.22 $\{X(s, t)\}$ is termed *stochastically increasing and directionally convex* in (s, t) in the sample-path sense, if for any four points $z^i = (s_i, t_i)$, $i = 1, 2, 3, 4$, as specified in Definition 2.20 (with (s_i, t_i) replacing (x_i, y_i)), there exist on a common probability space $(\Omega, \mathcal{F}, P)$ four random variables $\hat{X}_i$, $i = 1, 2, 3, 4$, such that $\hat{X}_i \overset{\mathrm{d}}{=} X(z^i)$, for all four i, and

$$
\begin{aligned}
\hat{X}_4(\omega) & \geq \max\{\hat{X}_2(\omega), \hat{X}_3(\omega)\}, \\
\hat{X}_1(\omega) + \hat{X}_4(\omega) & \geq \hat{X}_2(\omega) + \hat{X}_3(\omega),
\end{aligned}
$$

for all $\omega \in \Omega$.

Clearly, if $\{X(s,t)\}$ is stochastically increasing and directionally convex in (s,t) as defined earlier, then it is SICX-sp in s and in t and stochastically increasing and supermodular in (s,t) as defined in Definition 2.18. The converse, however, does not hold in general. Nevertheless, the converse does hold in a weaker sense, in the form of expectation.

Proposition 2.23 Suppose the family of random variables $\{X(s,t)\}$ is stochastically increasing and supermodular (resp. increasing and submodular) in (s,t) and stochastically increasing and convex (resp. increasing and concave) in s and in t, in any of the versions in Definitions 2.9, 2.10, and 2.11. Then, $\mathsf{E}h(X(s,t))$ is increasing and directionally convex (resp. directionally concave) in (s,t) for any increasing and convex (resp. concave) function $h(\cdot)$.

The preceding is immediate, taking into account the following, which can be directly verified.

Lemma 2.24 Suppose $\phi(x,y)$ is increasing and directionally convex (resp. directionally concave) in (x,y), and $h(z)$ is increasing and convex (resp. concave) in z. Then the composite function $h(\phi(x,y))$ is directionally convex (resp. directionally concave) in (x,y).

2.4   Markov Chain Applications

Let $\{X_n(s)\}$ be a discrete-time homogeneous Markov chain with state space on the nonnegative integers and initial state s, i.e., $X_0 = s$. Let

$$Y(x) \stackrel{\mathrm{d}}{=} [X_{n+1}|X_n = x].$$

That is, $Y(x)$ denote the generic random variable that has the same distribution as the position taken by the Markov chain in one transition starting from x. Not surprisingly, the spatial and temporal behavior of this homogeneous Markov chain is fully characterized by $Y(x)$. For example, clearly, if $Y(x) \geq x$, then X_n is stochastically increasing in n. If $\{Y(x) - x\}$ is stochastically increasing in x, then for any $s_1 \leq s_2$ (using "hat" to denote sample values):

$$
\begin{aligned}
X_0(s_1) + X_1(s_2) &\stackrel{\mathrm{d}}{=} s_1 + \hat{Y}(s_2) \\
&\geq s_2 + \hat{Y}(s_1) \\
&\stackrel{\mathrm{d}}{=} X_0(s_2) + X_1(s_1).
\end{aligned}
$$

Inductively, we can then show that $X_n(s)$ is supermodular in (n,s). Under the same conditions, we also have

$$\hat{X}_2 - \hat{X}_1 = \hat{Y}'(\hat{Y}(s)) - \hat{Y}(s) \geq \hat{Y}'(s) - s = \hat{X}_1 - \hat{X}_0,$$

where $\hat{Y}$ and $\hat{Y}'$ denote the sample values of the Y's that correspond to the two transitions, and the inequality is due to the increasingness of $Y(x) - x$, taking into account $\hat{Y}(s) \geq s$. Hence inductively, $\{X_n(s)\}$ satisfies SICX-sp in n (for any given s). Finally, it is similarly verified that $\{Y(x)\}$ satisfying SICX-sp implies $\{X_n(s)\}$ satisfying SICX-sp in s (for any given n). To summarize, we have the following:

Proposition 2.25 Suppose $\{Y(x)-x\}$ is stochastically increasing in x and $\{Y(x)\}$ satisfies SICX-sp. Then the Markov chain $\{X_n(s)\}$ satisfies SICX-sp in n and in s and is stochastically supermodular in (n, s). In particular, $\mathsf{E}[X_n(s)]$ is directionally increasing and convex in (n, s).

As an application of this, consider a continuous-time pure death process that starts at the initial state $X_0 = s$.

Proposition 2.26 $\{X_t(s)\}$ is a pure death process with initial state $X_0 = s$ and death rate $\mu(x)$ given $X_t = x$. Suppose $\mu(x)$ is increasing and concave in x, with $\mu(0) = 0$ and $\mu(x) > 0$ for all $x > 0$. Then, $X_t(s)$ is SICX-sp in s and in $-t$. In particular, $\mathsf{E}[X_t(s)]$ is directionally increasing and convex in $(s, -t)$ (i.e., increasing and convex in s, decreasing and convex in t, and submodular in (s, t)).

Proof. We use uniformization (refer to, e.g., [73]) to discretize time and generate the sample path of the pure death process. Let $\eta = \mu(s)$ be the uniformization constant. [Note $\eta \geq \mu(x)$ for all $x \leq s$, due to the increasingness of $\mu(x)$.] Let

$$0 = \tau_0 < \tau_1 < \ldots < \tau_n < \ldots$$

be a sequence of event epochs associated with the Poisson process with rate η. We will construct a discrete-time Markov chain $\{X_n(s)\}$, with initial state s, and use it to generate the path of the death process $\hat{X}_t$ by setting, for $n = 0, 1, 2, \ldots$,

$$\hat{X}_t = X_n, \quad t \in [\tau_n, \tau_{n+1}).$$

In order for the path generated to follow the same probability law as the original death process, the $Y(x)$ associated with $\{X_n(s)\}$ is specified as follows:

$$Y(x) = x - \mathbf{1}[\eta U \leq \mu(x)].$$

We want to show, following Proposition 2.25, that (i) $\{x - Y(x)\}$ is stochastically increasing in x and (ii) $\{Y(x)\}$ satisfies SICX-sp.

Because (i) follows immediately from the construction of Y, we show (ii). Pick x_i, $i = 1, 2, 3, 4$, such that

$$x_1 \leq x_2, x_3 \leq x_4; \quad x_1 + x_4 = x_2 + x_3.$$

Let

$$
\begin{aligned}
\hat{Y}_i &= x_i - \mathbf{1}[\eta U \le \mu(x_i)], \quad i = 1, 2, 4, \\
\hat{Y}_3 &= x_3 - \mathbf{1}[\eta U \le \mu(x_1)] \\
&\quad - \mathbf{1}[\mu(x_2) \le \eta U \le \mu(x_2) + \mu(x_3) - \mu(x_1)].
\end{aligned}
$$

Then, clearly, $\hat{Y}_i \overset{\mathrm{d}}{=} Y(x_i)$ for $i = 1, 2, 3, 4$ and $\hat{Y}_1, \hat{Y}_2, \hat{Y}_3 \le \hat{Y}_4$. Further, $\hat{Y}_1 + \hat{Y}_4 \ge \hat{Y}_2 + \hat{Y}_3$ is equivalent to

$$
\begin{aligned}
&\mathbf{1}[\eta U \le \mu(x_4)] \\
\le\; &\mathbf{1}[\eta U \le \mu(x_2)] + \mathbf{1}[\mu(x_2) \le \eta U \le \mu(x_2) + \mu(x_3) - \mu(x_1)] \\
=\; &\mathbf{1}[\eta U \le \mu(x_2) + \mu(x_3) - \mu(x_1)].
\end{aligned}
$$

But this follows immediately from the concavity of $\mu(x)$:

$$
\mu(x_1) + \mu(x_4) \le \mu(x_2) + \mu(x_3). \qquad \Box
$$

Corollary 2.27 Suppose $\{X_t(s)\}$ is a pure death process as in Proposition 2.26. Then $\{s - X_t(s)\}$ is a birth process that satisfies SICV-sp in s and in t and is stochastically supermodular in (s, t).

Proof. The birth process in question starts at 0, increases over time, eventually reaches s, and then stays at s. Its one-step transition is characterized by

$$
Y(x) = x + \mathbf{1}[\eta U \le \mu(s - x)].
$$

Similar to the proof of Proposition 2.26, we can show that $\{Y(x) - x\}$ is stochastically increasing in x, and $\{Y(x)\}$ satisfies SICV-sp. The desired properties follow, similar to those in Proposition 2.26. $\Box$

2.5   Notes

Stochastic and likelihood ratio orderings are common notions of stochastic orders; see, e.g., Ross [73] and Stoyan [90].

The coupling argument, used extensively in this chapter, plays a central role in the definitions of both stochastic convexity and submodularity and their proofs. It is built on the basic idea of constructing sample paths of stochastic processes under comparison on a common probability space and demonstrating that they possess certain desired relations. Refer to Kamae, Krengel, and O'Brien [48]. The sample-path version of stochastic convexity, SICX-sp in Definition 2.9 was originally developed in Shaked and Shanthikumar [80]; also see [81]; the strong version, SSICX in Definition 2.10 is due to Shanthikumar and Yao [83, 84]. The SSICX property holds in a wide range of stochastic models, including a variety of queueing networks, production systems under kanban control, and other discrete-event systems; refer to Glasserman and Yao [40] and [42].

The materials in §2.3 and §2.4 are drawn from Chang, Shanthikumar, and Yao [14]; also refer to Chang and Yao [15] and Shanthikumar and Yao [82]. The isotone property of a submodular function mentioned at the beginning of §2.3 is well known; refer to Topkis [98]. In some later chapters, we will need a strengthening of submodularity to what we call K-submodularity (Chapters 3 and 7).

3

Quality Control for Products with Warranty

The objective of this chapter is to develop an inspection procedure for end products that are supplied to customers with some type of warranty (or service contract), which obliges the manufacturer to provide repair, replacement, or, in some cases, refund to the consumer for a product that has failed within a certain period of time as specified by the contract.

Consider the context of batch manufacturing. Suppose we are dealing with a batch of N units and the quality of the batch is characterized by Θ, the proportion of defective units. Assume Θ is only known through its distribution. Both defective and nondefective units have random lifetimes with given distributions.

Suppose there is an inspection-repair procedure that can identify and repair all the defective units. Hence, if we follow a 100% inspection, we can guarantee that all N units are nondefective before they are shipped. However, inspection and repair do not come free. At the least, they will consume production capacity. The essence of the problem here is to strike a balance between the warranty cost and the inspection-repair cost.

The policy we shall identify and prove to be optimal has a simple, sequential structure: it is characterized by a sequence of threshold values, $d_{n_0} \leq \ldots \leq d_n \leq \ldots \leq d_{n_1}$, such that if D_n denotes the number of defective units among n inspected units, then the optimal policy is to stop inspection at the first n that satisfies $D_n < d_n$.

The key that underlies the optimality of this policy is a simple and intuitive monotone property: the higher the defective rate–not in terms of Θ, but in terms of its posterior estimate, given the outcome of the inspection– the more inspection an optimal policy will require. It turns out that this monotone property is a direct consequence of the warranty cost, as a function of the number of inspected units and the conditional defective rate, satisfying a so-called K-submodularity property, which is a strengthening of the usual notion of submodularity. Based on this property, we are able to identify several structural results of the optimal policy and eventually characterize the policy itself in terms of certain simple thresholds.

In §3.1 we spell out the precise details of the problem. We then elaborate on the K-submodularity property of the expected warranty cost in §3.2 and related properties of the conditional distribution of the defective rate in §3.3. The optimal control problem is formulated in §3.4, where several key structural properties of the optimal policy are established in Theorems 3.11, 3.12, and 3.14, which lead to a statement of the optimal policy in Theorem 3.16. A special case, the individual warranty model, is studied in §3.5. Two numerical examples and possible extensions are presented in §3.6.

3.1 Warranty Cost Functions

A batch of N units of a certain product has been completed on the production line. The units will supply customer demand, under some kind of warranty that will be specified later. We want to devise an inspection-repair procedure to ensure quality and to balance inspection-repair cost on the one hand and warranty cost on the other.

Assume each unit in the batch of N is either defective or nondefective. A nondefective unit has a lifetime of X, and a defective unit has a lifetime of Y. Both X and Y are random variables. Suppose X and Y are ordered under stochastic ordering, $X \geq_{st} Y$, i.e.,

$$P[X \geq a] \geq P[Y \geq a] \text{ for all } a \geq 0.$$

(Refer to Definition 2.1.)

Assume an inspection procedure can identify whether a unit is defective at a cost of c_i per unit. Each defective unit identified by the inspection is repaired, at a cost of c_r per unit, and becomes a nondefective unit.

The quality of the batch, before any inspection and repair, is represented by Θ, the proportion of defective units in the batch. Here Θ is assumed to be a random variable, with a known distribution function. Without loss of generality, assume $\Theta \in [\theta_0, \theta_1]$, where θ_0 and θ_1 are two given constants, $0 \leq \theta_0 \leq \theta_1 \leq 1$. (This essentially follows the quality model of Mamer [58].) Note that letting $\theta_0 = \theta_1 = \theta$ models the special case of a deterministic $\Theta \equiv \theta$. However, this special case restricts the number of defectives in n items to a binomial distribution, with a squared coefficient of variation equal to $(1 - \theta)/(n\theta)$, much too small–when n is large–for many applications.

Because there is no a priori discernible information about the quality of any units in the batch, we assume that each inspection will identify a defective unit with probability Θ and a nondefective unit with probability $1 - \Theta$. Note here that Θ is itself a random variable. That is, we do not have the exact information about the quality of the batch, although each inspection will improve our estimation of Θ, in the sense of obtaining an updated *conditional* distribution.

Let $Z(\theta)$ denote a random variable that is equal in distribution to Y (resp. X) with probability θ (resp. $1 - \theta$). That is, given $\Theta = \theta$, $Z(\theta)$ denotes the lifetime of a unit in the batch that has not been inspected.

Let $C(t)$ denote the warranty cost, a function of the lifetime of the units. (The functional form of $C(t)$ depends, of course, on the type of the warranty; see the examples later.) Specifically, suppose $\Theta = \theta$, and exactly n units in the batch are inspected (and the defectives repaired). Then, the expected warranty cost is

$$\phi(n, \theta) := \mathsf{E}[C(X_1 + \cdots + X_n + Z_{n+1}(\theta) + \cdots + Z_N(\theta))], \qquad (3.1)$$

where X_is and Z_js are i.i.d. random variables that follow the distributions of X and $Z(\theta)$, respectively. We assume that $C(t)$ is decreasing and convex in t (for the obvious motivation presented in the last section).

An example is in order. Consider the so-called "cumulative warranty" ([9]): it covers the batch as a whole, with a warranty period (for the entire batch) of NW time units, where W is a given positive constant. This type of warranty applies mostly to reliability systems, where spare parts (in cold stand-by) are used extensively. Let T denote the argument of $C(\cdot)$ in (3.1). Suppose the warranty cost takes the following form:

$$C(T) = (cN)[NW - T]^+/(NW) = c[N - T/W]^+, \qquad (3.2)$$

where $[x]^+$ denotes $\max\{x, 0\}$ and $c > 0$ is the selling price of each unit. Under this model, the manufacturer pays back part of the selling price on a pro rata basis. Here C is obviously a decreasing and convex function.

Another case of interest is when $C(\cdot)$ is an *additive* function. That is, (3.1) takes the following form:

$$\begin{aligned} \phi(n, \theta) &= \mathsf{E}[C(X_1)] + \cdots + \mathsf{E}[C(X_n)] + \mathsf{E}[C(Z_{n+1}(\theta))] \\ &\quad + \cdots + \mathsf{E}[C(Z_N(\theta))]. \end{aligned} \qquad (3.3)$$

This is the usual individual warranty model, i.e., the warranty applies to each individual unit instead of the batch as a whole. (For example, let $N = 1$ in (3.2) and apply the function to each individual unit.) In this case, we only need to assume the decreasing property of $C(\cdot)$, the convexity being replaced by the additivity.

In either case, the expected total (inspection, repair, and warranty) cost associated with a batch in which exactly $n \leq N$ units are inspected can be expressed as follows:

$$\Pi(n, \theta) = c_i n + c_r n\theta + \phi(n, \theta). \qquad (3.4)$$

In the rest of this section and the following three sections (§3.2 through §3.4) we will focus on the cost model in (3.1) and derive the optimal policy. In §3.5 we will show that the same optimal policy, in a simplified form, applies to the individual warranty model in (3.3) as well.

To conclude this section, we illustrate two points: (a) the ϕ function in (3.1) preserves the decreasing convexity of C, and (b) a relationship between the repair cost and the warranty cost is implicit in our model.

We will make frequent use of the following notation:

$$X_{1,n} := X_1 + \cdots + X_n, \quad Z_{n,N}(\theta) := Z_n(\theta) + \cdots + Z_N(\theta).$$

Lemma 3.1 Given θ, $\phi(n, \theta)$ is decreasing and convex in n.

Proof. Clearly, $X \geq_{st} Y$ implies $X \geq_{st} Z(\theta)$ (for any $\theta \in [\theta_0, \theta_1]$). For simplicity, we will omit the argument θ. Because the random variables are independent, we have

$$X_{1,n} + Z_{n+1,N} \leq_{st} X_{1,n+1} + Z_{n+2,N},$$

and hence

$$\mathsf{E}[C(X_{1,n} + Z_{n+1,N})] \geq \mathsf{E}[C(X_{1,n+1} + Z_{n+2,N})],$$

because $C(t)$ is a decreasing function. That is, ϕ is decreasing in n.

To establish convexity, we use coupling, similar to the approach in Examples 2.16 and 2.17. Because $X \geq_{st} Z$, we can have, for $j = 1, 2$, $X^j \geq Z^j$ almost surely (a.s.), with X^j and Z^j equal in distribution to X and Z, respectively. Let $\tau := X_{1,n-1} + Z_{n+2,N}$. Maintain independence wherever necessary. Because $C(t)$ is convex, we have

$$C(\tau + X^1 + X^2) + C(\tau + Z^1 + Z^2) \geq C(\tau + X^1 + Z^2) + C(\tau + X^2 + Z^1) \quad \text{a.s.}$$

Taking expectations on both sides yields

$$\phi(n+1) + \phi(n-1) \geq \phi(n) + \phi(n) = 2\phi(n),$$

which is the required convexity. □

Clearly, in this proof, we have actually established that the warranty cost $C(X_{1,n} + Z_{n+1,N})$, as a function of n, is SICX-sp, as in Definition 2.9, which is stronger than the convexity of the expected cost.

In the earlier model description, we assumed that each defective unit identified by the inspection must be repaired. Although this assumption appears reasonable and innocuous, it does impose certain restrictions on the warranty cost. If the warranty cost were sufficiently low or the repair cost relatively high, then the manufacturer might choose to repair only some of the defective units while taking a chance on the others. Hence, in order to be consistent with our assumption that all defective units are repaired, we will insist that the following condition be satisfied:

$$c_r \leq \mathsf{E}[C(X_{1,n-1} + Y + Z_{n+1,N}(\theta)] - \mathsf{E}[C(X_{1,n-1} + X + Z_{n+1,N}(\theta)]$$

for all $n \leq N - 1$ and all θ. The condition says that if we have identified a certain defective unit in the batch, it pays to have it repaired, because the repair cost plus the subsequent warranty cost will not exceed the warranty cost with the defective unit shipped unrepaired. In Lemma 3.5 we will show that the right side in the preceding inequality is decreasing in n. Hence, the condition reduces to a single inequality (the case of $n = N - 1$):

$$c_r \leq \mathsf{E}[C(X_{1,N-1} + Y)] - \mathsf{E}[C(X_{1,N})]. \tag{3.5}$$

Throughout our study of the cumulative warranty model (from here through §3.4), we will assume that condition in (3.5) is always in force.

3.2 K-Submodularity

When the defective rate Θ is a known constant, say θ, the quality control problem becomes a *static* optimization problem: we want to find the optimal $n^*(\theta)$ that minimizes the expected total cost $\Pi(n, \theta)$ in (3.4), for a given θ. (The problem is static, because there is nothing to be gained after each inspection, in terms of estimating the defective rate.)

Specifically, we want to establish the following monotone property of the optimal solution: $n^*(\theta') \geq n^*(\theta)$ for all $\theta' \geq \theta$. That is, the lower the quality of the batch (in terms of a larger θ), the more we need to inspect. It turns out that the key to this is the notion of K-submodularity defined soon.

Recall the isotone property involved in minimizing a submodular function (§2.3): suppose $x^*(y)$ is the optimal solution to the minimization problem, $\min_x g(x, y)$, for a given y; then $x^*(y)$ is increasing in y. However, here we are interested in a slightly different problem: $\min_x[Kxy + g(x, y)]$, where $K > 0$ is a constant [cf. $\Pi(n, \theta)$ in (3.4)]. Because Kxy is a *super*modular function, the submodularity of g will not guarantee the increasingness of the optimal solution $x^*(y)$ in y. In order to maintain the isotone property of the optimal solution, we need to strengthen the submodularity of g.

Definition 3.2 A bivariate function, $g(x, y)$, is called K-*submodular*, if for some $K \geq 0$, we have

$$[g(x_1, y_2) + g(x_2, y_1)] - [g(x_1, y_1) + g(x_2, y_2)] \geq K(x_1 - x_2)(y_1 - y_2)$$

for all $x_1 \geq x_2$ and $y_1 \geq y_2$.

Remark 3.3 (1) Following Definition 3.2, a K-submodular function has the following geometric property: consider its values on the four corner points (x_1, y_1), (x_1, y_2), (x_2, y_1), and (x_2, y_2) of a rectangle; the off-diagonal sum is greater than the diagonal sum by at least K times the area of the rectangle. Obviously, K-submodularity specializes to submodularity with $K = 0$.

(2) The term, K-submodularity, is inspired by the notion of K-convexity, which plays a key role in proving the optimality of (s, S) inventory policies; refer to Scarf [77].

Clearly, from Definition 3.2, $g(x, y)$ is K-submodular if and only if $Kxy + g(x, y)$ is submodular. Hence, the next lemma follows from the known isotone property in minimizing a submodular function mentioned earlier. However, we still give a proof, because we need the details to support the later extension of the result to the stochastic setting (see Proposition 3.8).

Lemma 3.4 Let $x^*(y)$ be an optimal solution to $\min_x [Kxy + g(x, y)]$, for a given y. Then, $x^*(y)$ is increasing in y, if $g(x, y)$ is K-submodular. (In the event of multiple optimal solutions, then $x^*(y)$ is taken to be the largest one.)

Proof. Denote $x_1 = x^*(y)$. Pick $y' > y$. We show that if $x_2 < x_1$, then x_2 cannot provide a better solution than x_1 at y', hence x_2 cannot be $x^*(y')$. Use contradiction. Suppose x_2 yields a better solution at y', i.e.,

$$Kx_2y' + g(x_2, y') < Kx_1y' + g(x_1, y').$$

From the optimality of x_1 at y, we have

$$Kx_1y + g(x_1, y) \leq Kx_2y + g(x_2, y).$$

Summing up the two inequalities, we have

$$[g(x_1, y) + g(x_2, y')] - [g(x_1, y') + g(x_2, y)] < K(x_1 - x_2)(y' - y),$$

contradicting the K-submodularity of g. $\square$

Lemma 3.5 The expected warranty cost $\phi(n, \theta)$ is K-submodular in (n, θ) with $K = c_r$, i.e.,

$$[\phi(n, \theta) + \phi(n - 1, \theta')] - [\phi(n, \theta') + \phi(n - 1, \theta)] \geq c_r(\theta' - \theta)$$

for all n and all $\theta' > \theta$.

Proof. Consider $\phi(n - 1, \theta) - \phi(n, \theta)$. Conditioning on $Z_n(\theta) = X$ or Y, we have

$$
\begin{aligned}
&\phi(n - 1, \theta) - \phi(n, \theta) \\
=\ &\theta\{\mathsf{E}[C(X_{1,n-1} + Y + Z_{n+1,N}(\theta))] - \mathsf{E}[C(X_{1,n-1} + X + Z_{n+1,N}(\theta))]\} \\
:=\ &\theta\Delta(n, \theta). \tag{3.6}
\end{aligned}
$$

We want to show $\theta' \Delta(n, \theta') - \theta \Delta(n, \theta) \geq c_r(\theta' - \theta)$ for all $n \leq N$ and all $\theta' > \theta$.

We first note that $\Delta(n, \theta)$ is increasing in θ. This follows easily from the definition of Δ in (3.6) and a coupling argument (similar to the one in the proof of Lemma 3.1), taking into account that (a) $X \geq_{st} Y$, (b) $Z(\theta)$ is (stochastically) decreasing in θ, and (c) $C(t)$ is decreasing and convex. Next, note that the decreasing convexity of ϕ in n (Lemma 3.1) implies that $\Delta(n, \theta)$ is decreasing in n [cf. (3.6)].

Making use of the two preceding properties, we have

$$\theta' \Delta(n, \theta') - \theta \Delta(n, \theta)$$
$$\geq \quad (\theta' - \theta) \Delta(n, \theta)$$
$$\geq \quad (\theta' - \theta) \Delta(N, \theta)$$
$$= \quad (\theta' - \theta) \{ \mathsf{E}[C(X_{1,N-1} + Y)] - \mathsf{E}[C(X_{1,N})] \}$$
$$\geq \quad (\theta' - \theta) c_r,$$

where the last inequality follows from (3.5). $\square$

Clearly, $c_i n + \phi(n, \theta)$ is also K-submodular. From Lemma 3.4, we have the following.

Proposition 3.6 The optimal solution $n^*(\theta)$ that solves $\min_{0 \leq n \leq N} \Pi(n, \theta)$ for any given θ is increasing in θ.

Next, suppose instead of $\Pi(n, \theta)$, we want to minimize

$$\bar{\Pi}(n, \Theta) := \mathsf{E}[\Pi(n, \Theta)] = c_i n + c_r n \mathsf{E}[\Theta] + \mathsf{E}[\phi(n, \Theta)]. \qquad (3.7)$$

(Note that this is still a static optimization problem: the defective rate, although a random variable, does not change with respect to n.)

Lemma 3.5 can be readily adapted to the stochastic setting; refer to §2.3, Definition 2.18 in particular. From Lemma 3.5, we know that $\phi(n - 1, \theta) - \phi(n, \theta) - c_r \theta$ is increasing in θ, and hence, we know the following.

Proposition 3.7 The warranty cost $\phi(n, \Theta)$ is stochastically K-submodular, with $K = c_r$, in the following sense:

$$\{ \mathsf{E}[\phi(n - 1, \Theta')] + \mathsf{E}[\phi(n, \Theta)] \} - \{ \mathsf{E}[\phi(n - 1, \Theta)] + \mathsf{E}[\phi(n, \Theta')] \} \geq c_r \mathsf{E}[\Theta' - \Theta]$$

for all n and $\Theta' \geq_{st} \Theta$. $\square$

Making use of this inequality and mimicking the proof of Lemma 3.4, we have the following.

Proposition 3.8 Let $n^*(\Theta)$ be the solution to $\min_{0 \leq n \leq N} \bar{\Pi}(n, \Theta)$ for a given Θ. Then,
(i) $\Theta' \geq_{st} \Theta$ implies $n^*(\Theta') \geq n^*(\Theta)$; and in particular,
(ii) $n_0^* \leq n^*(\Theta) \leq n_1^*$, where for $j = 0, 1$, n_j^* is the optimal solution to $\min_{0 \leq n \leq N} \Pi(n, \theta_j)$.

3.3 Conditional Distribution for Defectives

Let D_n denote the number of defectives uncovered through inspecting n units. We are interested in the conditional distribution of Θ given $D_n = d$. Let $\Theta_n(d) := [\Theta | D_n = d]$. This is the quantity that embodies the sequential nature of the original quality control problem. Therefore, to make preparations for deriving the optimal policy in the next section, we establish, in the next two lemmas, (a) the monotone properties of $\Theta_n(d)$ with respect to n and d and (b) K-submodularity properties similar to Proposition 3.7, with Θ replaced by $\Theta_n(d)$.

We want to show that the likelihood ratio ordering applies to $\Theta_n(d)$, as n and d vary. This then implies (following Lemma 2.3) the stochastic ordering, which, although weaker, is in this case more cumbersome to prove directly.) For convenience, assume Θ has a density function $f_\Theta(x)$, and denote the density function of $\Theta_n(d)$ as $f_{\Theta_n(d)}(x)$. Then, we have

$$
\begin{aligned}
f_{\Theta_n(d)}(x) &= \mathsf{P}[\Theta \in dx | D_n = d] \\
&= \frac{\mathsf{P}[D_n = d | \Theta = x] f_\Theta(x)}{\int_\theta \mathsf{P}[D_n = d | \Theta = \theta] f_\Theta(\theta) d\theta} \\
&= \frac{x^d (1-x)^{n-d} f_\Theta(x)}{\mathsf{E}[\Theta^d (1-\Theta)^{n-d}]}, \quad x \in [\theta_0, \theta_1]. \tag{3.8}
\end{aligned}
$$

Lemma 3.9 For all n and $d \le n$, we have

$$
\Theta_n(d+1) \ge_{\mathrm{lr}} \Theta_{n+1}(d+1) \ge_{\mathrm{lr}} \Theta_n(d) \ge_{\mathrm{lr}} \Theta_{n+1}(d).
$$

In particular, $\Theta_n(d)$ is increasing in d and decreasing in n, both in the sense of the likelihood ratio ordering.

Proof. We prove the third inequality; the other two are similarly proved. From (3.8), we have, for all $x \ge y$,

$$
\begin{aligned}
\frac{f_{\Theta_n(d)}(x)}{f_{\Theta_{n+1}(d)}(x)} &= \frac{1}{1-x} \cdot \frac{\mathsf{E}[\Theta^d(1-\Theta)^{n+1-d}]}{\mathsf{E}[\Theta^d(1-\Theta)^{n-d}]} \\
&\ge \frac{1}{1-y} \cdot \frac{\mathsf{E}[\Theta^d(1-\Theta)^{n+1-d}]}{\mathsf{E}[\Theta^d(1-\Theta)^{n-d}]} = \frac{f_{\Theta_n(d)}(y)}{f_{\Theta_{n+1}(d)}(y)},
\end{aligned}
$$

hence the desired likelihood ratio ordering. $\square$

Lemma 3.10 For all $n < N$ and $d < n$, we have

$$
\begin{aligned}
c_r \mathsf{E}[\Theta_n(d)] &+ \mathsf{E}[\phi(n+1, \Theta_n(d))] - \mathsf{E}[\phi(n, \Theta_n(d))] \\
&\ge c_r \mathsf{E}[\Theta_n(d+1)] + \mathsf{E}[\phi(n+1, \Theta_n(d+1))] \\
&\quad - \mathsf{E}[\phi(n, \Theta_n(d+1))] \tag{3.9}
\end{aligned}
$$

and

$$c_r \mathsf{E}[\Theta_n(d)] + \mathsf{E}[\phi(n+1, \Theta_n(d))] - \mathsf{E}[\phi(n, \Theta_n(d))]$$
$$\geq c_r \mathsf{E}[\Theta_{n-1}(d)] + \mathsf{E}[\phi(n, \Theta_{n-1}(d))]$$
$$-\mathsf{E}[\phi(n-1, \Theta_{n-1}(d))]. \tag{3.10}$$

Proof. Because $\Theta_n(d+1) \geq_{st} \Theta_n(d)$, which follows from the likelihood ratio ordering established in Lemma 3.9, the inequality in (3.9) follows from Proposition 3.7.

Similarly, to prove (3.10), from Lemma 3.9, we have $\Theta_{n-1}(d) \geq_{st} \Theta_n(d)$, and hence

$$c_r \mathsf{E}[\Theta_n(d)] + \mathsf{E}[\phi(n+1, \Theta_n(d))] - \mathsf{E}[\phi(n, \Theta_n(d))]$$
$$\geq c_r \mathsf{E}[\Theta_{n-1}(d)] + \mathsf{E}[\phi(n+1, \Theta_{n-1}(d))] - \mathsf{E}[\phi(n, \Theta_{n-1}(d))]$$
$$\geq c_r \mathsf{E}[\Theta_{n-1}(d)] + \mathsf{E}[\phi(n, \Theta_{n-1}(d))] - \mathsf{E}[\phi(n-1, \Theta_{n-1}(d))],$$

where the first inequality makes use of Proposition 3.7 and the second makes use of the decreasing convexity of ϕ in n (Lemma 3.1). $\square$

3.4 Optimal Policy

Let $V_n(d)$ be the expected total remaining cost, following an optimal policy, after n units are inspected and d units are found defective. Then the optimal cost for the original problem is $V_0(0)$.

Let

$$\Phi_n(d) := \mathsf{E}[\phi(n, \Theta_n(d))]$$

and

$$\Psi_n(d) := c_i + [c_r + V_{n+1}(d+1)]\mathsf{P}[D_{n+1} = d+1 | D_n = d]$$
$$+ V_{n+1}(d)\mathsf{P}[D_{n+1} = d | D_n = d].$$

Clearly, $\Phi_n(d)$ and $\Psi_n(d)$ represent the expected cost associated with the two actions we can take in stage n and state d: either stop inspection (i.e., ship the batch without inspecting the remaining $N - n$ units) or continue inspecting one more unit.

Hence, we have the following recursion:

$$V_n(d) = \min\{\Phi_n(d), \Psi_n(d)\}$$

for $0 \leq n \leq N - 1$ and $V_N(d) = \Phi_N(d)$. Furthermore, from standard results in dynamic programming, (e.g., Ross [74], chapter 1), we know that the optimal policy that minimizes $V_0(0)$ has the following general structure: at each stage n, stop in state d if $\Phi_n(d) < \Psi_n(d)$; continue inspecting more units if $\Phi_n(d) > \Psi_n(d)$; and choose either action when $\Phi_n(d) = \Psi_n(d)$. We will reveal more structural properties of the optimal policy and eventually establish its threshold nature.

Observing that

$$P[D_{n+1} = d + 1 | D_n = d] = E[\Theta_n(d)] = \frac{E[\Theta^{d+1}(1 - \Theta)^{n-d}]}{E[\Theta^d(1 - \Theta)^{n-d}]},$$

we can also express $\Psi_n(d)$ as

$$\Psi_n(d) = c_i + [c_r + V_{n+1}(d + 1)]E[\Theta_n(d)] + V_{n+1}(d)(1 - E[\Theta_n(d)]).$$

Yet another expression for $\Psi_n(d)$, which will be used later, is:

$$\Psi_n(d) = c_i + c_r E[\Theta_n(d)] + E[V_{n+1}(D_{n+1})|D_n = d]. \tag{3.11}$$

For each $0 \le n \le N - 1$, define

$$\mathbf{S}_n := \{d : 0 \le d \le n, \Phi_n(d) \le \Psi_n(d)\},$$

$$\overline{\mathbf{S}}_n := \{d : 0 \le d \le n, \Phi_n(d) \ge \Psi_n(d)\}.$$

That is, $\mathbf{S}_n$ is the set of states in which it is optimal to stop (after inspecting n units), while $\overline{\mathbf{S}}_n$ is the set of states in which it is optimal to continue the inspection. (Note that here $\overline{\mathbf{S}}_n$ is not just the complement of $\mathbf{S}_n$.)

In what follows, we present the structures of the optimal policy in Theorems 3.11 through 3.14, which lead to a statement of the optimal policy in Theorem 3.16.

The first theorem specifies the range for the number of inspected units: the optimal policy must inspect a minimum of n_0^* units and a maximum of n_1^* units, where n_0^* and n_1^* are characterized in Proposition 3.8.

Theorem 3.11 For each n and all $0 \le d \le n$, (i) $d \in \overline{\mathbf{S}}_n$ if $n < n_0^*$, and (ii) $d \in \mathbf{S}_n$ if $n \ge n_1^*$.

Proof. Based on Lemma 3.1 and Proposition 3.6, we know that $\Pi(n, \theta)$ is convex in n and reaches its minimum at $n^*(\theta)$, which falls between n_0^* and n_1^*. Hence, it is decreasing in n for $n < n_0^*$, i.e., $\Pi(n + 1, \theta) \le \Pi(n, \theta)$, for $n < n_0^*$. Hence, replacing θ by $\Theta_n(d)$ and taking expectations, we have $\overline{\Pi}(n + 1, \Theta_n(d)) \le \overline{\Pi}(n, \Theta_n(d))$. That is, from (3.7),

$$c_i(n + 1) + c_r(n + 1)E[\Theta_n(d)] + E[\phi(n + 1, \Theta_n(d))]$$
$$\le \quad c_i n + c_r n E[\Theta_n(d)] + E[\phi(n, \Theta_n(d))],$$

which simplifies to

$$c_i + c_r E[\Theta_n(d)] + E[\phi(n + 1, \Theta_n(d))] \le E[\phi(n, \Theta_n(d))] = \Phi_n(d).$$

Observing that

$$
\begin{aligned}
E[\phi(n+1, \Theta_n(d))] &= E[\phi(n+1, \Theta)|D_n = d] \\
&\geq E[V_{n+1}(D_{n+1})|D_n = d],
\end{aligned}
$$

where the inequality is implied by the optimality of V_{n+1}, we have

$$
\begin{aligned}
\Psi_n(d) &= c_i + c_r E[\Theta_n(d)] + E[V_{n+1}(D_{n+1})|D_n = d] \\
&\leq c_i + c_r E[\Theta_n(d)] + E[\phi(n+1, \Theta_n(d))] \leq \Phi_n(d).
\end{aligned}
$$

Hence, $d \in \overline{\mathbf{S}}_n$, when $n < n_0^*$.

To show that $d \in \mathbf{S}_n$, for $n \geq n_1^*$, we use induction. Clearly, $V_N(d) = \Phi_N(d)$, i.e., $d \in \mathbf{S}_N$. Suppose $d \in \mathbf{S}_{n+1}$, i.e., $V_{n+1}(d) = \Phi_{n+1}(d)$ for any $d \leq n+1$. We then have

$$
\begin{aligned}
\Psi_n(d) &= c_i + c_r E[\Theta_n(d)] + E[V_{n+1}(D_{n+1})|D_n = d] \\
&= c_i + c_r E[\Theta_n(d)] + E[\phi(n+1, \Theta_{n+1}(D_{n+1}))|D_n = d] \\
&= c_i + c_r E[\Theta_n(d)] + E[\phi(n+1, \Theta_n(d))] \geq \Phi_n(d),
\end{aligned}
$$

where the inequality follows from $\bar{\Pi}(n+1, \Theta_n(d)) \geq \bar{\Pi}(n, \Theta_n(d))$, for $n \geq n_1^*$, which in turn follows from the fact that $\Pi(n, \theta)$ is increasing in n for $n \geq n_1^*$ (because $\Pi(n, \theta)$ is convex in n and reaches its minimum at $n^*(\theta) \leq n_1^*$). This implies $d \in \mathbf{S}_n$. $\square$

The next theorem establishes the following monotone property of the optimal policy: at each stage n, if it is optimal to continue inspection in state d, then it is also optimal to continue in state $d+1$. (This follows intuitively from Lemma 3.9, because state $d+1$ implies a poorer quality, in terms of an increased estimate of the defective rate.) This monotone property then leads to a threshold structure of the optimal policy: once the number of defectives identified, D_n, exceeds a threshold, it is optimal to continue with more inspections.

Theorem 3.12 For each n: $n_0^* \leq n < n_1^*$, and all $d < n$, if $d \in \overline{\mathbf{S}}_n$, then $(d+1) \in \overline{\mathbf{S}}_n$.

Proof. We want to show that for each n: $n_0^* \leq n < n_1^*$, and each $d < n$,

$$
\Psi_n(d) \leq \Phi_n(d) \quad \Rightarrow \quad \Psi_n(d+1) \leq \Phi_n(d+1).
$$

Instead, we prove via induction a stronger result: that $\Psi_n(d) - \Phi_n(d)$ is decreasing in d, for each given n.

When $n = n_1^* - 1$, since $(n_1^*, d) \in \mathbf{S}_n$, we have

$$
E[V_{n+1}(D_{n+1})|D_n] = E[\phi(n+1, \Theta_n(d))]
$$

in (3.11). Hence,

$$
\Psi_n(d) - \Phi_n(d) = c_i + c_r E[\Theta_n(d)] + E[\phi(n+1, \Theta_n(d))] - E[\phi(n, \Theta_n(d))],
$$

which is decreasing in d, following (3.9). Consequently, when $n = n_1^* - 1$, $V_n(d) = \Phi_n(d)$ for $d < d^*$, and $V_n(d) = \Psi_n(d) \leq \Phi_n(d)$ for $d \geq d^*$, with d^* defined as

$$d^* := \min\{d \leq n : \psi_n(d) \leq \phi_n(d)\}.$$

Hence, $V_n(d) - \Phi_n(d)$ is also decreasing in d.

Next, consider $n < n_1^* - 1$. As induction hypothesis, assume that $V_{n+1}(d) - \Phi_{n+1}(d)$ is decreasing in d. Then

$$
\begin{aligned}
&\Psi_n(d) - \Phi_n(d) \\
=\ & c_i + c_r \mathsf{E}[\Theta_n(d)] + \mathsf{E}[V_{n+1}(D_{n+1})|D_n = d] - \mathsf{E}[\phi(n, \Theta_n(d))] \\
=\ & \{c_i + c_r \mathsf{E}[\Theta_n(d)] + \mathsf{E}[\phi(n+1, \Theta_n(d))] - \mathsf{E}[\phi(n, \Theta_n(d))]\} \\
+\ & \{\mathsf{E}[V_{n+1}(D_{n+1})|D_n = d] - \mathsf{E}[\phi(n+1, \Theta_n(d))]\}. \qquad (3.12)
\end{aligned}
$$

It remains to show that the two parts (in braces) in (3.12) are both decreasing in d. This is obvious for the first part, following (3.9). For the second part, note that the two terms can be written as

$$\mathsf{E}[V_{n+1}(d + I_n(d)) - \Phi_{n+1}(d + I_n(d))|D_n = d], \qquad (3.13)$$

where $I_n(d)$ is a 0-1 binary random variable that equals 1 with probability $\mathsf{E}[\Theta_n(d)]$, and hence is stochastically increasing in d (Lemma 3.9). Hence, the required decreasing (in d) property follows from the induction hypothesis, which implies that $V_{n+1}(d) - \Phi_{n+1}(d)$ is decreasing in d. $\square$

Corollary 3.13 For each n: $n_0^* \leq n < n_1^*$, let

$$d_n := \min\{d \leq n : \Psi_n(d) \leq \Phi_n(d)\}.$$

Then d_n is well defined, with $d \in \mathbf{S}_n$ for $d < d_n$, and $d \in \overline{\mathbf{S}}_n$ for $d_n \leq d \leq n$.

Complementing Theorem 3.12, we next show the other half of the monotone property of the optimal policy: if it is optimal to stop at stage n in state $D_n = d$, then it is also optimal to stop at stage $n + 1$ in the same state d. Consequently, the threshold values d_n in Corollary 3.13 must be increasing in n.

Theorem 3.14 For each n: $n_0^* \leq n < n_1^*$, and all $d \leq n$, if $d \in \mathbf{S}_n$ then $d \in \mathbf{S}_{n+1}$.

Proof. Similar to the proof of Theorem 3.12, here it suffices to show that $\Psi_n(d) - \Phi_n(d)$ is increasing in n, i.e.,

$$\psi_n(d) - \phi_n(d) \leq \psi_{n+1}(d) - \phi_{n+1}(d) \qquad (3.14)$$

for any given $d \leq n$, $n_0^* \leq n < n_1^*$. The induction steps are exactly the same, except here we use (3.10) instead of (3.9). In particular, the first part in (3.12) is increasing in n, following (3.10). To show that (3.13) is also increasing in n, note that it dominates, via the induction hypothesis, $E[V_n(d + I_n(d)) - \Phi_n(d + I_n(d))|D_n = d]$, which, in turn, dominates

$$E[V_n(d + I_{n-1}(d)) - \Phi_n(d + I_{n-1}(d))|D_n = d],$$

because $I_n(d)$ is decreasing in n and $V_n(d) - \Phi_n(d)$ is decreasing in d. This yields the desired increasing (in n) property. $\square$

Because $\Psi_n(d) - \Phi_n(d)$ is increasing in n, $d < d_n$ will always imply $d < d_{n+1}$ for any d. Therefore, we have the following.

Corollary 3.15 The d_n values in Corollary 3.13 are increasing in n, i.e., $d_n \leq d_{n+1}$.

Summarizing the results in Theorems 3.11, 3.12 and 3.14, we have the following.

Theorem 3.16 The optimal policy that minimizes $V_0(0)$ is to start from inspecting n_0^* units, continue to inspect one more unit at a time, and stop as soon as the total number of inspected units n satisfies: $D_n < d_n$ or when $n = n_1^*$.

3.5   The Individual Warranty Model

We now return to the individual warranty model introduced in §3.1. Recall in this case that the cost function $C(\cdot)$ is additive. From (3.3), we have

$$\phi(n, \theta) = NE[C(X)] + N\theta(E[C(Y)] - E[C(X)]) - n\theta(E[C(Y)] - E[C(X)]).$$

(Note here that we do not need to assume the convexity of $C(\cdot)$. The decreasingness of $C(\cdot)$, however, is still needed. This ensures that

$$E[C(Y)] - E[C(X)] \geq 0$$

for $X \geq_{st} Y$.) Consequently,

$$\begin{aligned} \Pi(n, \theta) = \ & NE[C(X)] + N\theta(E[C(Y)] - E[C(X)]) \\ & + c_i n - n\theta(E[C(Y)] - E[C(X)] - c_r). \end{aligned} \quad (3.15)$$

When $E[C(Y)] - E[C(X)] - c_r > 0$, define

$$\hat{\theta} := \frac{c_i}{E[C(Y)] - E[C(X)] - c_r},$$

which guarantees that $\Pi(n, \theta)$ is increasing in n for $\theta \leq \hat{\theta}$ and decreasing in n for $\theta \geq \hat{\theta}$.

Lemma 3.17 Let n_θ^* be the optimal solution for $\min_{0 \le n \le N} \Pi(n, \theta)$ of (3.15). Then
(i) $n_\theta^* = 0$ for any θ when $c_r \ge \mathsf{E}[C(Y)] - \mathsf{E}[C(X)]$;
(ii) $n_\theta^* = 0$ when $c_r < \mathsf{E}[C(Y)] - \mathsf{E}[C(X)]$ and $\theta \le \hat{\theta}$; and
(iii) $n_\theta^* = N$ when $c_r < \mathsf{E}[C(Y)] - \mathsf{E}[C(X)]$ and $\theta > \hat{\theta}$.

Proof. $\Pi(n, \theta)$ is increasing in n under the conditions in (i) and (ii) and decreasing in n under the conditions in (iii). $\square$

Note that while the condition $c_r < \mathsf{E}[C(Y)] - \mathsf{E}[C(X)]$ in (ii) and (iii) is what (3.5) specializes to here, the condition in (i) goes in the opposite direction. Regardless, because $\Pi(n, \theta)$ here is linear in n, in all three cases, there is a meaningful solution. Specifically, provided that $\Theta \equiv \theta$ is a known constant, the optimal (static) policy is to either inspect all units in the batch or not inspect at all, depending on the relationship between the cost and quality data in question. This is consistent with the recommendation of Tapiero and Lee in [94], as well as what is usually followed in practice.

Intuitively, Lemma 3.17 recommends that if the repair cost is high ($c_r \ge \mathsf{E}[C(Y)] - \mathsf{E}[C(X)]$) or if the defective rate is low ($\theta \le \hat{\theta}$), then do no inspection at all; otherwise (i.e., if the repair cost is low *and* the defective rate is high), do 100% inspection. Note the quantity $\hat{\theta}$ plays the role of a threshold for the defective rate.

Next consider the original problem with Θ being a random variable and the optimal policy that minimizes $V_0(0)$. From (3.15), obviously $\Pi(n, \theta)$ is linear in n and linear in θ. It is also easy to see that when $c_r \le \mathsf{E}[C(Y)] - \mathsf{E}[C(X)]$, $\Pi(n, \theta)$ is submodular in (n, θ). Consequently, $\bar{\Pi}(n, \Theta)$ is linear in n, linear in $\mathsf{E}[\Theta]$, and submodular in $(n, \mathsf{E}[\Theta])$ if $c_r \le \mathsf{E}[C(Y)] - \mathsf{E}[C(X)]$. Hence, Proposition 3.8 and Theorem 3.11 also hold here. These, along with Lemma 3.17, lead to the following.

Proposition 3.18 (i) When $c_r \ge \mathsf{E}[C(Y)] - \mathsf{E}[C(X)]$, then $n_0^* = n_1^* = 0$, and the optimal policy is not to inspect any unit.
(ii) When $c_r < \mathsf{E}[C(Y)] - \mathsf{E}[C(X)]$, there are two cases:

(a) if $\theta_0 \ge \hat{\theta}$, then $n_0^* = n_1^* = N$, and the optimal policy is to inspect all N units; and

(b) if $\theta_1 \le \hat{\theta}$, then $n_0^* = n_1^* = 0$, and the optimal policy is not to inspect any unit.

Note that $\theta_0 \ge \hat{\theta}$ implies $\Theta \ge \hat{\theta}$ and $\theta_1 \le \hat{\theta}$ implies $\Theta \le \hat{\theta}$. Hence, Proposition 3.18 confirms that the optimal rules for a constant Θ in Lemma 3.17 are also optimal in the general setting of a random Θ.

Now, what remains is the most interesting case of $\theta_0 < \hat{\theta} < \theta_1$ under $c_r < \mathsf{E}[C(Y)] - \mathsf{E}[C(X)]$. First note that $n_0^* = 0$ and $n_1^* = N$ in this case, following Lemma 3.17. Hence, in principle all N units could be subject to inspection. We can do better, however, in particular, improving on n_0^*.

Lemma 3.19 (i) Consider the case of $c_r < E[C(Y)] - E[C(X)]$ and $\hat{\theta} \in (\theta_0, \theta_1)$. After inspecting n units, if $E[\Theta_n(d)] \geq \hat{\theta}$, then it is optimal to continue inspection.

(ii) Denote $\hat{n} := \max\{n : 0 \leq n \leq N, E[\Theta_n(0)] \geq \hat{\theta}\}$, with the understanding that $\hat{n} = -1$ if the set is empty. Then it is optimal to inspect at least $n^* := \hat{n} + 1$ units.

Proof. Note that

$$\Phi_n(d) = N E[C(X)] + (N - n) E[\Theta_n(d)](E[C(Y)] - E[C(X)])$$

and conditioning on the quality of the $(n + 1)^{st}$ unit, we have

$$E[\Phi_{n+1}(D_{n+1})|D_n = d] = \Phi_n(d) - E[\Theta_n(d)](E[C(Y)] - E[C(X)])$$

or

$$\Phi_n(d) = E[\Theta_n(d)](E[C(Y)] - E[C(X)]) + E[\Phi_{n+1}(D_{n+1})|D_n = d].$$

On the other hand,

$$\begin{aligned} \Psi_n(d) &= c_i + c_r E[\Theta_n(d)] + E[V_{n+1}(D_{n+1})|D_n = d] \\ &\leq c_i + c_r E[\Theta_n(d)] + E[\Phi_{n+1}(D_{n+1})|D_n = d], \end{aligned}$$

where the inequality is from the definition of $V_{n+1}(\cdot)$. If $E[\Theta_n(d)] \geq \hat{\theta}$, we have

$$c_i + c_r E[\Theta_n(d)] \leq E[\Theta_n(d)](E[C(Y)] - E[C(X)]).$$

Therefore, $\Psi_n(d) \leq \Phi_n(d)$, i.e., it is optimal to continue inspection if $E[\Theta_n(d)] \geq \hat{\theta}$.

Note that $d = 0$ when the inspection starts, and $E[\Theta_n(0)]$ is decreasing in n, from Lemma 3.9. Because, for any n, $n \leq \hat{n}$ implies $E[\Theta_n(0)] \geq \hat{\theta}$ and it is optimal to continue inspection, at least $\hat{n} + 1$ units should be inspected. This is the conclusion in (ii). $\square$

For n with $n^* \leq n < N$, the optimal decision follows the sequence of thresholds in Corollary 3.13. To summarize, we have the following.

Proposition 3.20 In the case of $c_r < E[C(Y)] - E[C(X)]$ and $\hat{\theta} \in (\theta_0, \theta_1)$, the optimal policy is to start from inspecting a sample of n^* units (where n^* follows the definition in Lemma 3.19), and then to continue with one unit at a time and stop as soon as the total number of inspected units n satisfies $D_n < d_n$, or $n = N$. $\square$

Because here the only requirement for the cost function $C(\cdot)$ is the decreasing property, specializing the function in different ways, we can model a wide range of individual warranty types, e.g., those with free replacement or rebate replacement (e.g., [9], or Lie and Chun [56]).

3.6   Examples and Extensions

Consider a batch of $N = 30$ units. Suppose the defective rate Θ is uniformly distributed between $\theta_0 = 5\%$ and $\theta_1 = 30\%$. The lifetime of a good unit, X, is uniformly distributed on $[70, 110]$; and the lifetime of a defective unit, Y, is uniformly distributed on $[30, 70]$. The inspection and repair costs are $c_i = 0.5$ and $c_r = 1$. Suppose there is a cumulative, pro rata rebate warranty of the type in (3.2) associated with the batch, with the unit price $c = 100$ and a warranty period $W = 82$ per unit.

Following Theorem 3.16, the following thresholds can be computed:

$$d_n = 0, \ n \le 8; \quad d_9 = d_{10} = d_{11} = 1; \quad d_{12} = d_{13} = d_{14} = 2;$$

$$d_{15} = d_{16} = 3; \quad d_{17} = 4; \quad d_{18} = 5; \quad d_{19} = 6; \quad d_{20} = 8; \quad d_{21} = 12.$$

These thresholds guide the inspection at each step. For example, as long as there is at least one defect in 11 inspected units, inspection should continue; whereas inspection can be terminated if there are fewer than 4 defectives in 17 inspected units. Under no circumstances should inspection continue beyond $n_1 = 22$ inspected units ($n_0 = 0$).

Following this optimal policy, the expected number of inspected units is 14.79, uncovering an average of 2.76 defective units. The total expected cost is 12.964. In contrast, the expected costs under zero inspection and full inspection policies are 36.41 and 20.25, respectively.

As a second example, consider the individual warranty model. Use the preceding data and apply the pro rata rebate warranty to each individual unit. The only change is to set the unit price at $c = 11$. The computed thresholds are:

$$d_n = 0, \ n \le 7; \quad d_n = 1, \ 8 \le n \le 12; \quad d_n = 2, \ 13 \le n \le 18;$$

$$d_n = 3, \ 19 \le n \le 23; \quad d_n = 4, \ 24 \le n \le 28; \quad d_{29} = 5.$$

(Here, $n_0 = 0$ and $n_1 = 30$.) This optimal policy yields an expected number of 20.73 inspected units, with an average of 4.01 identified defective units. The total expected cost is 26.65. This is much closer (than the case of cumulative warranty) to the expected costs under zero inspection and full inspection policies: 28.51 and 27.49, respectively. (When the unit price is $c = 100$, this falls into the special case of Proposition 3.18 (i), and the optimal policy is to do full inspection.)

Imperfect inspection and/or repair can be easily incorporated into our model. The only change is to modify the distributions of X and Y, so that each inspected (and repaired) unit will have a lifetime that is a mixture of X and Y. This will not affect the structure of the model or the form of the optimal policy established earlier. (Note here that we assume an imperfect inspection will only identify a defective unit with a certain probability, but will never mistake nondefective units as defective. Similarly, we assume an imperfect repair will only transform a defective unit into a nondefective unit with a certain probability but will never make a nondefective unit defective. If these assumptions do not hold, obviously the optimal policy will be structurally different. For example, it might become justified *not* to repair a unit that is identified as defective by the inspection.)

The imperfection of inspection and repair will be reflected in the threshold values. For example, if the inspection and repair are very ineffective (in identifying and correcting defective units), then the threshold value n_0^* will be low, while d_ns will be high, so that the optimal policy will stop inspection early, or conduct no inspection at all.

The random variables, X and Y, can be replaced by random vectors without affecting any structure of the optimal policy. The random vectors can model, for example, the so-called two-dimensional warranties in [87]. That is, in addition to lifetime, there is a second dimension that explicitly accounts for usage (e.g., six years or 600,000 miles).

3.7 Notes

In the literature of statistical quality control (or statistical process control), there are several approaches to quality assurance through inspection and repair. The simplest is to do some back-of-envelope calculations based on the given cost data and the *average* defective rate ($\mathsf{E}[\Theta]$) and then choose from two actions: either do 100% inspection on the batch or do no inspection at all. A better, but more involved approach is to inspect a small sample (of size n, say), and if the number of defectives identified exceeds the expected value, $n\mathsf{E}[\Theta]$ by a certain number of "sigma" (standard deviation), then inspect all the remaining units; otherwise, stop inspection and accept the batch. The most sophisticated existing approach is perhaps the CUSUM technique (refer to, e.g., Thompson and Koronacki [97], chapter 4): inspect the units one at a time; at each step update the cumulative sum of the log-likelihood estimate of the defective rate; continue if the sum falls within a prespecified interval, say $[\alpha, \beta]$; stop inspection and ship the whole batch if the sum falls below α; and inspect all the remaining units if the sum exceeds β.

Common to all these approaches is that the *policy* is prespecified; the issue then becomes essentially a parametric design problem of finding one or two threshold values: the breakeven point in the back-of-envelope analysis, the upper limit on the acceptable number of defective units along with the sample size in the acceptance-rejection approach, and the α and β values in the CUSUM technique.

In contrast, our focus here (also refer to [21]) is on identifying a *policy* that can be proven optimal instead of merely finding optimal parameters of prespecified policies. Our approach is based on dynamic programming or, more specifically, sequential analysis. The basic idea of sequential analysis (Wald [100]), in our context, calls for identifying a procedure (i.e., policy) by which items are inspected one by one, and each time after inspecting a unit, we decide to either stop inspection or continue, depending on the outcome of the inspection up to that point.

There is an extensive body of literature that studies the statistical, social-economical, and behavioral aspects of warranty; see, e.g., the survey articles of Blischke [9] and Singpurwalla and Wilson [87]. There are also studies on various optimization issues that arise in warranties and related services, e.g., Djamaludi, Murthy, and Wilson [32]; Mamer [58]; Murthy and Nguyen [62]; Murthy, Wilson, and Djamaludi [63]; Nguyen and Murthy [64]; and Thomas [96]. The model studied in Tapiero and Lee [94] is quite similar to the individual warranty model in §3.5. For the static optimization problem, i.e., with $\Theta \equiv \theta$, a given constant, Tapiero and Lee demonstrated that the optimal policy is either 0% or 100% inspection. They also pointed out that when Θ is a random variable, the optimal policy will be different from these extreme-point rules. This optimal policy is completely characterized in §3.5 here, as a special case of our general model.

4
Process Control in Batch Production

The topic of this chapter is the control of a machining process through inspecting the units it has produced. Today's advanced sensor technology notwithstanding, it often remains prohibitively expensive or infeasible to do online direct inspection. This is because it is very difficult to develop a sensor that is highly sensitive to the parameters it is supposed to monitor while simultaneously insensitive to other parameters and rugged enough to withstand the harsh environment in which it is used; refer to [23]. An indirect off-line inspection remains the predominant practical means for process control and revision.

In general terms, our problem can be described as follows. A machine produces a batch of units over each time period. At the end of each period, the units are inspected (assuming, for the time being, that inspection is cost free). The outcome of the inspection, along with other quality data and system parameters, is used to determine whether the machining process has shifted away from the desired in control state. If it is decided that the state is 'out of control', then a process 'revision' is ordered. This usually includes maintenance and recalibration of the machine or more elaborate repair work. Otherwise, the process is allowed to continue into the next period. In the presence of inspection cost, there is also the need to decide whether to inspect each batch produced.

We develop a Markov decision process approach for the process control problem, by which both machine revision and inspection decisions are made in each period, based on the information obtained from all previous periods. To the extent that we use the probability of the machine being out of control as the state variable, as the machine state itself is not directly observable, we are effectively dealing with a partially observable Markov decision process. We prove that the optimal control for machine revision is of threshold types, with or without inspection cost; whereas the optimal control of inspection has a more involved structure characterized by multiple threshold values.

We start with a description of the problem in §4.1 and then we derive several preliminary results. We then present background materials on MDP and formulate our problem as an MDP in §4.2. In §4.3, we study the discounted-cost model, analyze the problem structure, and prove the optimality of threshold-type policies. We then present results for the average-cost model in §4.4. In §4.5, we consider the special case of no inspection cost, where we relate the discounted-cost and average-cost models through certain monotone properties of the optimal thresholds.

4.1 Machine Revision

Suppose the machine is always in one of two states: in control or out of control, denoted as state 0 and state 1, respectively. The duration of staying in state 0 is exponentially distributed with parameter λ. Decisions are made periodically at fixed constant time intervals, or 'periods'. In each period, a batch of N items is produced by the machine, where N is a positive integer. If the machine is in state 0 during the period, the probability for each item to be defective—or, the defective rate—is Θ_0. If the machine is in state 1, the defective rate is Θ_1. Here Θ_0 and Θ_1 are random variables with known distributions, independent of each other. We will assume that, within a batch, the items are produced one by one, so that if the machine state shifts during the period, then some items in the batch—those produced before the shift—will have defective rate Θ_0, and others will have defective rate Θ_1. (Note this includes the following as a special case: all the items in the batch are produced all together, so that the defective rate of the whole batch is Θ_1 if the machine state shifts during the period.)

Assume that the defective rate is larger when the machine is out of control, i.e., $\Theta_1 \geq \Theta_0$ (with probability 1). (This will hold, in particular, if Θ_0 and Θ_1 belong to two nonoverlapping intervals of the [0,1] continuum. For example, the defective rate is below 5% when the machine is in control and above 25% when the machine is out of control.) Although it is possible to envision other weaker orderings in certain applications, they will not be sufficient to lead to the kind of threshold structure of the optimal policy that we shall focus on later.

Let the length of each period be the time unit of choice, and refer to the interval $[i-1, i)$ as the ith period. At the beginning of each period, we decide whether to revise the machine, based on the system history up to that point. If we choose to revise the machine, we pay a cost C_R, and the machine will be set to the in control state immediately.

If we decide not to revise the machine, the machine state will remain unknown, and if it is out of control, it will stay that way. At the end of the period, a batch of N units is produced. We then decide whether to inspect the batch. If the decision is to do inspection, we pay an inspection cost C_I for the entire batch. The inspection identifies all defective units, which are all repaired, at a cost of C_D per unit. (Assume the inspection-repair process takes no time.) If the decision is not to inspect, there is a penalty cost C_P for each undetected defective unit. Because passing a defective unit to the downstream is usually more costly than getting it repaired right away, it is natural to assume $C_P \geq C_D$. The process then enters into the next period, and the steps are repeated.

Note that by convention, we always view the inspection decision as taking place at the *end* of each period, while the revision decision takes place at the *beginning* of the *next* period. Because inspection consumes no time, each revision decision follows immediately. However, the quality information obtained from the inspection is useful in making the revision decision.

Our objective is to find an optimal policy in revising the machine and inspecting the products so that the total discounted or average-cost is minimized.

For $i = 1, 2, \cdots$, let D_i denote the number of defectives in the ith batch (i.e., the one that is produced in the ith period). Let (a_i, b_i) denote the pair of control actions in period i, where $a_i = 1, 0$ denotes the actions of revising and not revising the machine, and $b_i = 1, 0$ denotes the actions of inspecting and not inspecting the batch produced. Figure 4.1 depicts the sequence of decisions carried out over time, along with the production process.

From the preceding problem specification, the possible actions in each period are $(1, 1)$, $(1, 0)$, $(0, 1)$, and $(0, 0)$. Let

$$H_i = (a_1, b_1, \mathcal{D}_1; a_2, b_2, \mathcal{D}_2; \cdots; a_i, b_i, \mathcal{D}_i)$$

denote the system history up to (the end of) the ith period. Here and later, $\mathcal{D}_j$ denotes a sample realization from D_j if $b_j = 1$; it is void if $b_j = 0$. Denote A_i (resp. A_i^c) as the event that the machine is out of control (resp. in control) by the end of period i, *after* the inspection (if any) has taken place. Let Y_0 denote the probability that the machine is out of control at time 0; and for any $i \geq 1$, let $Y_i := \mathsf{P}[A_i | Y_0, H_i]$, i.e., Y_i is the posterior probability of the machine being out of control at the end of the ith period, given Y_0 and the history H_i. Note that as a conditional probability, Y_i is itself a random variable.

Let $p_0(k)$ (resp. $p_1(k)$) denote the probability that given the machine is in control (resp. out of control) in a period, there are k defective items produced in the period; and let $p_2(k)$ denote the same probability given

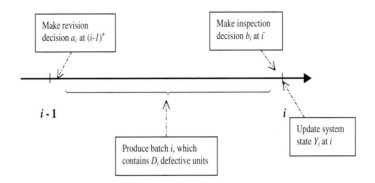

FIGURE 4.1. Timeline of making a decision.

the machine state has shifted in the period. Then

$$
\begin{aligned}
p_0(k) &= \int_\theta \binom{N}{k} \theta^k (1-\theta)^{N-k} dF_{\Theta_0}(\theta) \\
&= \binom{N}{k} E[\Theta_0^k (1-\Theta_0)^{N-k}],
\end{aligned}
\tag{4.1}
$$

where $F_{\Theta_0}(\cdot)$ is the distribution function of Θ_0. Similarly,

$$
p_1(k) = \binom{N}{k} E[\Theta_1^k (1-\Theta_1)^{N-k}].
\tag{4.2}
$$

To describe $p_2(k)$, let D denote the number of defectives in a batch when the machine state shifts (from in control to out of control during the period); let $\mathcal{E}_n$ denote the event that the machine shifts its state when it is processing the nth unit of the batch; and let T denote the in control duration. Then

$$
p_2(k) = \sum_{n=1}^{N} P[D = k \,|\, \mathcal{E}_n] P[\mathcal{E}_n]
\tag{4.3}
$$

with

$$
P[\mathcal{E}_n] = P[\frac{n-1}{N} \leq T < \frac{n}{N} \,|\, T < 1] = \frac{e^{-\frac{n-1}{N}\lambda} - e^{-\frac{n}{N}\lambda}}{1 - e^{-\lambda}}
$$

and

$$P[D = k \,|\, \mathcal{E}_n] \;=\; \sum_{j=0}^{n-1} \binom{n-1}{j} \mathsf{E}[\Theta_0^j (1 - \Theta_0)^{n-1-j}]$$
$$\cdot \binom{N - n + 1}{k - j} \mathsf{E}[\Theta_1^{k-j}(1 - \Theta_1)^{(N-n+1)-(k-j)}].$$

Lemma 4.1 If $a_{i-1} = 0$, then

$$
\begin{aligned}
Y_i &= f_0(Y_{i-1}) \\
&:= Y_{i-1} + (1 - Y_{i-1})(1 - e^{-\lambda}),
\end{aligned}
\tag{4.4}
$$

when $b_{i-1} = 0$, and

$$
\begin{aligned}
Y_i &= f_1(Y_{i-1}, \mathcal{D}_i) \\
&:= \frac{Y_{i-1}p_1(\mathcal{D}_i) + (1 - Y_{i-1})p_2(\mathcal{D}_i)(1 - e^{-\lambda})}{Y_{i-1}p_1(\mathcal{D}_i) + (1 - Y_{i-1})[p_2(\mathcal{D}_i)(1 - e^{-\lambda}) + p_0(\mathcal{D}_i)e^{-\lambda}]}
\end{aligned}
\tag{4.5}
$$

when $b_{i-1} = 1$. If $a_{i-1} = 1$, then Y_i follows the expressions with $Y_{i-1} = 0$.

Proof. We only need to prove the case of $a_{i-1} = 0$ (no revision). The other case, $a_{i-1} = 1$, follows from the fact that the machine is always returned to the in control state after revision.

Note that (4.4) follows immediately from conditioning on whether the machine state shifts before $i - 1$ or after $i - 1$ but before i.

To prove (4.5), suppose $j < i - 1$ is the time of the last revision before $i - 1$. Let $j = 0$ if no revision was ever made before $i - 1$. Then clearly, Y_i only depends on Y_j (the first state after the revision) and the history $(\mathcal{D}_{j+1}, ..., \mathcal{D}_i)$, which we will denote as $\mathcal{D}_{j+1,i}$. Hence,

$$
\begin{aligned}
Y_i &= \mathsf{P}[A_i \,|\, Y_j, \mathcal{D}_{j+1,i}] \\
&= 1 - \mathsf{P}[A_i^c \,|\, Y_j, \mathcal{D}_{j+1,i}] \\
&= 1 - \frac{\mathsf{P}[A_i^c, \mathcal{D}_{j+1,i} \,|\, Y_j]}{\mathsf{P}[\mathcal{D}_{j+1,i} \,|\, Y_j]}.
\end{aligned}
\tag{4.6}
$$

For simplicity, we omit Y_j from the expressions, with the understanding that all of the following probabilities are conditioning on Y_j. We have

$$
\begin{aligned}
&\mathsf{P}[A_{i-1}, \mathcal{D}_{j+1,i}] \\
&= \mathsf{P}[\mathcal{D}_{j+1,i-1}]\mathsf{P}[A_{i-1} \,|\, \mathcal{D}_{j+1,i-1}]\mathsf{P}[\mathcal{D}_i \,|\, \mathcal{D}_{j+1,i-1}, A_{i-1}] \\
&= \mathsf{P}[\mathcal{D}_{j+1,i-1}]Y_{i-1}p_1(\mathcal{D}_i),
\end{aligned}
$$

$$
\begin{aligned}
&\mathsf{P}[A_{i-1}^c, \mathcal{D}_{j+1,i}] \\
&= \mathsf{P}[\mathcal{D}_{j+1,i-1}]\mathsf{P}[A_{i-1}^c \,|\, \mathcal{D}_{j+1,i-1}]\mathsf{P}[\mathcal{D}_i \,|\, \mathcal{D}_{j+1,i-1}, A_{i-1}^c] \\
&= \mathsf{P}[\mathcal{D}_{j+1,i-1}](1 - Y_{i-1})\mathsf{P}[\mathcal{D}_i \,|\, A_{i-1}^c] \\
&= \mathsf{P}[\mathcal{D}_{j+1,i-1}](1 - Y_{i-1})[p_2(\mathcal{D}_i)(1 - e^{-\lambda}) + p_0(\mathcal{D}_i)e^{-\lambda}],
\end{aligned}
$$

where the last equality is obtained by conditioning on whether the state shifts in period i. Hence,

$$\begin{aligned} &P[\mathcal{D}_{j+1,i}] \\ = \ &P[\mathcal{D}_{j+1,i-1}]\{Y_{i-1}p_1(\mathcal{D}_i) \\ &+(1-Y_{i-1})[p_2(\mathcal{D}_i)(1-e^{-\lambda})+p_0(\mathcal{D}_i)e^{-\lambda}]\}. \end{aligned} \tag{4.7}$$

Similarly,

$$P[\mathcal{D}_{j+1,i}, A_i^c] = P[\mathcal{D}_{j+1,i-1}](1-Y_{i-1})e^{-\lambda}p_0(\mathcal{D}_i). \tag{4.8}$$

Substituting (4.7) and (4.8) into (4.6) yields the desired expression in (4.5).
□

4.2   MDP Formulation

4.2.1   MDP Essentials

Lemma 4.1 enables us to formulate the problem as a discrete-time Markov decision process (MDP) with a *countable* state space.

Briefly, a discrete-time MDP works as follows: there is a state space, denoted S; and for each state $y \in S$, $A(y)$ is a set of available control actions. In state $y \in S$, if $\mathbf{a} \in A(y)$ is chosen, then a cost $r(y, \mathbf{a})$ is incurred, and the system will transit to another state $x \in S$ with probability $p_{yx}(\mathbf{a})$.

A policy $\pi = \{\pi_0, \pi_1, \cdots\}$ is a sequence of rules, where π_t specifies the action to be taken at the tth period (decision epoch). In general, π_t may depend on the time epoch t and on the system history (up to t), and it can also take a randomized form by associating the actions with probabilities. A policy is said to be *Markovian* if π_t, for any $t \geq 0$, is independent of the history (before t). Furthermore, a Markovian policy is said to be stationary if it is time-invariant in the sense that the action it takes at any decision epoch depends only on the system state but *not* on the time. Hence, for a stationary policy π, we can omit the time index and write $\pi(z)$ to denote the action it takes in state z.

For a discount factor α, with $0 < \alpha < 1$, starting from the initial state y, the long-run discounted cost objective under policy π is

$$V_{\pi,\alpha}(y) = \mathsf{E}_\pi\Big[\sum_{t=0}^{\infty} \alpha^t r(Y_t, \mathbf{a}_t) \,|\, Y_0 = y\Big], \tag{4.9}$$

where E_π is the expectation with respect to the probability measure corresponding to policy π; and Y_t and $\mathbf{a}_t$ are, respectively, the state and the action taken in period t.

A policy π^* is said to be optimal (for a given α) if

$$V_{\pi^*,\alpha}(y) = \inf_\pi V_{\pi,\alpha}(y) \qquad \forall y \in S.$$

Similarly, starting from state y, the long-run average-cost objective is:

$$\bar{V}_\pi(y) = \varliminf \frac{\mathsf{E}_\pi[\sum_{t=0}^n r(Y_t, \mathbf{a}_t) \mid Y_0 = y]}{n+1}; \qquad (4.10)$$

and a policy π^* is said to be optimal if

$$\bar{V}_{\pi^*}(y) = \inf_\pi \bar{V}_\pi(y) \qquad \forall y \in S.$$

Let

$$V_\alpha(y) = \inf_\pi V_{\pi,\alpha}(y) \qquad \text{and} \qquad \bar{V}(y) = \inf_\pi \bar{V}_\pi(y)$$

for any $y \in S$. i.e., $V_\alpha(y)$ is the optimal (α-) discounted-cost function and $\bar{V}(y)$ is the optimal average-cost function.

Suppose $r(\cdot, \cdot)$ is bounded, S is countable, and $A(y)$ is finite or compact for any $y \in S$. The preliminary results summarized in Lemmas 4.2-4.4 are standard; refer to, for example, Puterman [71] or Ross [74].

Lemma 4.2 Consider the MDP specified earlier.

(i) $V_\alpha(y)$ is the unique solution to the following discounted optimal equation:

$$V_\alpha(y) = \min_{\mathbf{a} \in A(y)} \{r(y, \mathbf{a}) + \alpha \sum_x p_{yx}(\mathbf{a})V_\alpha(x)\}. \qquad (4.11)$$

(ii) The policy that selects the action that minimizes the right-hand side of (4.11) is optimal.

(iii) Let $V_\alpha^0(y) = 0$ and

$$V_\alpha^{m+1}(y) = \min_{\mathbf{a} \in A(y)} \{r(y, \mathbf{a}) + \alpha \sum_x p_{yx}(\mathbf{a})V_\alpha^m(x)\}$$

for $m \geq 0$. Then,

$$\lim_{m \to \infty} V_\alpha^m(y) = V_\alpha(y) \qquad \text{for any } y \in S.$$

Lemma 4.3 If there exists a bounded function $h(y)$ and a constant g such that

$$g + h(y) = \min_{\mathbf{a} \in A(y)} \{r(y, \mathbf{a}) + \sum_x p_{yx}(\mathbf{a})h(x)\}, \qquad (4.12)$$

then the policy π^* that prescribes the action that maximizes the right-hand side of (4.12) is optimal under the average-cost criterion, with

$$g = \bar{V}_{\pi^*}(y) = \inf_\pi \bar{V}_\pi(y) \qquad \forall x \in S.$$

Here (4.12) is generally called the average optimal equation for the discrete-time MDP.

Lemma 4.4 Suppose there exist a state $y_0 \in S$ and $\mathcal{K} < \infty$ such that

$$|V_\alpha(y) - V_\alpha(y_0)| < \mathcal{K}$$

for all α and all $y \in S$, then the $h(y)$ and g in Lemma 4.3 exist, with

$$g = \lim_{\alpha \to 1} (1 - \alpha) V_\alpha(y_0)$$

and

$$h(y) = \lim_{n \to \infty} [V_{\alpha_n}(y) - V_{\alpha_n}(y_0)]$$

for some sequence $\alpha_n \to 1$.

4.2.2   The MDP Problem

To formulate our problem with the MDP models described earlier, let $\mathbf{S}_0 = \{0\}$; and for $i \geq 1$, let

$$\mathbf{S}_i^0 = \{x : x = f_0(y), \ y \in \mathbf{S}_{i-1}\},$$

$$\mathbf{S}_i^1 = \{x : x = f_1(y, k), \ y \in \mathbf{S}_{i-1}, k = 0, 1, ..., n\},$$

and $\mathbf{S}_i = \mathbf{S}_i^0 \cup \mathbf{S}_i^1$. Define the state space as $\mathbf{S} = \bigcup_{i=0}^{\infty} \mathbf{S}_i$, which is obviously countable, and assume that the initial state $Y_0 \in \mathbf{S}$. (For example, a special case is to assume $Y_0 = 0$, i.e., initially the machine is in control. Because every time a revision takes place, the machine state is returned to zero, the in control state, even this stronger assumption does not lose much generality.)

As mentioned earlier, in each state $y \in S$, the possible actions are $(1, 1)$, $(1, 0)$, $(0, 1)$ and $(0, 0)$, and the transition probabilities follow Lemma 4.1, with $k = 0, 1, ..., n$:

$$\begin{aligned} &\mathsf{P}[Y_i = f_1(y, k) \,|\, Y_{i-1} = y, a_{i-1} = 0, b_{i-1} = 1] \\ =\ & y p_1(k) + (1 - y)[p_2(k)(1 - e^{-\lambda}) + p_0(k)e^{-\lambda}], \end{aligned}$$

$$\begin{aligned} &\mathsf{P}[Y_i = f_1(0, k) \,|\, Y_{i-1} = y, a_{i-1} = 1, b_{i-1} = 1] \\ =\ & p_2(k)(1 - e^{-\lambda}) + p_0(k)e^{-\lambda}, \end{aligned}$$

and

$$\begin{aligned} &\mathsf{P}[Y_i = f_0(y) \,|\, Y_{i-1} = y, a_{i-1} = 0, b_{i-1} = 0] \\ =\ & \mathsf{P}[Y_i = f_0(0) \,|\, Y_{i-1} = y, a_{i-1} = 1, b_{i-1} = 0] \\ =\ & 1. \end{aligned}$$

Recall that the cost for each identified and unidentified defective unit is C_D and C_P, respectively. Let $r(y, b)$ denote the expected cost for defectives in period i, given $Y_{i-1} = y$, $a_{i-1} = 0$, and $b_{i-1} = b$ for $b = 1, 0$. Conditioning on the time when the machine state shifts, we can derive

$$
\begin{aligned}
r(y, b) &= C(b)[N\mathsf{E}(\Theta_1)]\mathsf{P}[A_{i-1} \mid Y_{i-1} = y] \\
&\quad + C(b)\{[N\mathsf{E}(\Theta_0)]\mathsf{P}[A_i^c \mid Y_{i-1} = y] \\
&\quad + \sum_{j=1}^{N}[(j-1)\mathsf{E}(\Theta_0) + (N-j+1)\mathsf{E}(\Theta_1)]\mathsf{P}[\mathcal{E}_j \mid Y_{i-1} = y]\} \\
&= yC(b)[N\mathsf{E}(\Theta_1)] + (1-y)C(b)\{ N\mathsf{E}(\Theta_0)e^{-\lambda} \\
&\quad + \sum_{j=1}^{N}[(j-1)\mathsf{E}(\Theta_0) + (N-j+1)\mathsf{E}(\Theta_1)] \\
&\quad \cdot [\exp(-\frac{j-1}{N}\lambda) - \exp(-\frac{j}{N}\lambda)] \}
\end{aligned}
\tag{4.13}
$$

for $b = 1, 0$, with $C(1) := C_D$ and $C(0) := C_P$. Recall $\mathcal{E}_j$ is defined in (4.3).

Lemma 4.5 (i) $r(y, b)$ is linear and increasing in y, for $b = 1, 0$.
(ii) $r(y, 0) - r(y, 1)$ is also linear and increasing in y.

Proof. (i) Linearity is obvious from (4.13). For increasingness, the summation part in (4.13) is dominated by $N\mathsf{E}(\Theta_1)(1 - e^{-\lambda})$, because $\mathsf{E}(\Theta_0) \leq \mathsf{E}(\Theta_1)$, Hence, the coefficient of $-y$ is dominated by

$$
C(b)[N\mathsf{E}(\Theta_0)e^{-\lambda} + N\mathsf{E}(\Theta_1)(1 - e^{-\lambda})],
$$

which in turn is dominated by $C(b)[N\mathsf{E}(\Theta_1)]$, the coefficient of y.
(ii) Note that $r(y, 0) - r(y, 1)$ is equal to the expression in (4.13) with $C(b)$ substituted by $C_P - C_D$. Because $C_D \leq C_P$, we can follow the same argument that proves the result in (i). (ii) is obtained by following a discussion similar to proof of (i). $\square$

4.3 Discounted-Cost Model

4.3.1 Optimality Equations

The following lemma is a direct application of Lemma 4.2, taking into account that in our problem there are four possible action pairs in each period.

Lemma 4.6 (i) The optimal discounted-cost function $V_\alpha(y)$ is the unique solution to the following equation:

$$
\begin{aligned}
V_\alpha(y) = \min \{ &C_R + C_I + r(0, 1) + \alpha\mathsf{E}[V_\alpha(f_1(0, D))], \\
&C_R + r(0, 0) + \alpha V_\alpha(f_0(0)), \\
&C_I + r(y, 1) + \alpha\mathsf{E}[V_\alpha(f_1(y, D))], \\
&r(y, 0) + \alpha V_\alpha(f_0(y)) \}.
\end{aligned}
\tag{4.14}
$$

(ii) The policy that prescribes actions to minimize the right-hand side of equation (4.14) is a stationary optimal policy.

(iii) Define $V_\alpha^0(y) = 0$, and

$$
\begin{aligned}
V_\alpha^m(y) \;=\; & \min \{ C_R + C_I + r(0,1) + \alpha E[V_\alpha^{m-1}(f_1(0,D))], \\
& C_R + r(0,0) + \alpha V_\alpha^{m-1}(f_0(0)), \\
& C_I + r(y,1) + \alpha E[V_\alpha^{m-1}(f_1(y,D))], \\
& r(y,0) + \alpha V_\alpha^{m-1}(f_0(y)) \}.
\end{aligned} \tag{4.15}
$$

Then, $\lim_{m\to\infty} V_\alpha^m(y) = V_\alpha(y)$ for any initial state y . □

Define

$$
V_\alpha(1,y) = \min\{C_R + V_\alpha(2,0), V_\alpha(2,y)\} \tag{4.16}
$$

and

$$
\begin{aligned}
V_\alpha(2,y) \;=\; & \min\{C_I + r(y,1) + \alpha E[V_\alpha(1,f_1(y,D))], \\
& r(y,0) + \alpha V_\alpha(1,f_0(y))\}.
\end{aligned} \tag{4.17}
$$

First, we want to show that there exists a unique solution to these equations, i.e., $V_\alpha(1,y)$ and $V_\alpha(2,y)$ are well defined.

Define a set of recursive equations as follows: $V_\alpha^0(1,y) = 0$, $V_\alpha^0(2,y) = 0$,

$$
V_\alpha^m(1,y) = \min\{C_R + V_\alpha^m(2,0), V_\alpha^m(2,y)\}, \tag{4.18}
$$

and

$$
\begin{aligned}
V_\alpha^m(2,y) \;=\; & \min\{C_I + r(y,1) + \alpha E[V_\alpha^{m-1}(1,f_1(y,D))], \\
& r(y,0) + \alpha V_\alpha^{m-1}(1,f_0(y))\}.
\end{aligned} \tag{4.19}
$$

Lemma 4.7 The following limits exist: $\lim_{m\to\infty} V_\alpha^m(1,y) = V_\alpha(1,y)$ and $\lim_{m\to\infty} V_\alpha^m(2,y) = V_\alpha(2,y)$; and they are the unique solution to the equations in (4.16) and (4.17). Furthermore, $V_\alpha(1,y) = V_\alpha(y)$, where $V_\alpha(y)$ is the solution to (4.14), and hence the optimal-cost function.

Proof. Substituting (4.19) into (4.18), we obtain

$$
\begin{aligned}
V_\alpha^m(1,y) \;=\; & \min\{C_R + C_I + r(0,1) + \alpha E[V_\alpha^{m-1}(1,f_1(0,D))], \\
& C_R + r(0,0) + \alpha V_\alpha^{m-1}(1,f_0(0)), \\
& C_I + r(y,1) + \alpha E[V_\alpha^{m-1}(1,f_1(y,D))], \\
& r(y,0) + \alpha V_\alpha^{m-1}(1,f_0(y))\}.
\end{aligned}
$$

From Lemma 4.6(iii), $\lim_{m\to\infty} V_\alpha^m(1,y)$ exists and is the unique solution to equation (4.14). Denote this limit as $V_\alpha(1,y)$. Letting $m \to \infty$ in (4.19), we have

$$
\begin{aligned}
& \lim_{m\to\infty} V_\alpha^m(2,y) \\
=\; & \min\{C_I + r(y,1) + \alpha E[V_\alpha(1,f_1(y,D))], \\
& r(y,0) + \alpha V_\alpha(1,f_0(y))\}.
\end{aligned}
$$

Hence, the limit $\lim_{m\to\infty} V_\alpha^m(2,y)$, denoted $V_\alpha(2,y)$, exists and is the unique solution to the equation in (4.17). This, in turn, implies that

$$\lim_{m\to\infty} V_\alpha^m(1,y) := V_\alpha(1,y)$$

is the unique solution to (4.16). Because both $V_\alpha(1,y)$ and $V_\alpha(y)$ are solutions to (4.14), by Lemma 4.6 (i), we have $V_\alpha(1,y) = V_\alpha(y)$. $\square$

Lemma 4.7 enables us to compute the optimal discounted-cost function for the original problem through (4.16) and (4.17). Next, we show that the stationary optimal policy itself can also be obtained from (4.16) and (4.17). Define a stationary policy π_α through (4.16) and (4.17) as follows: in period i, if the system is in state y, choose $a_i = 1$ if and only if

$$C_R + V_\alpha(2,0) \leq V_\alpha(2,y);$$

furthermore, if $a_i = 1$ is chosen, then choose $b_i = 1$ if and only if

$$C_I + r(0,1) + \alpha E[V_\alpha(1, f_1(0,D))] \leq r(0,0) + \alpha V_\alpha(1, f_0(0)), \qquad (4.20)$$

and if $a_i = 0$ is chosen, then choose $b_i = 1$ if and only if

$$C_I + r(y,1) + \alpha E[V_\alpha(1, f_1(y,D))] \leq r(y,0) + \alpha V_\alpha(1, f_0(y)). \qquad (4.21)$$

Then, it is easy to verify that π_α always chooses control actions that minimize the right-hand side of equation (4.14). From Lemma 4.6, any policy that chooses control actions in this manner is optimal. Therefore, we have the following.

Theorem 4.8 The stationary policy π_α as specified previously is optimal; in particular, $V_\alpha(\pi_\alpha, y) = V_\alpha(y)$, for any state $y \in S$. $\square$

4.3.2 Structural Properties

Here we show that both $V_\alpha(1,y)$ and $V_\alpha(2,y)$ are increasing and concave in y (See Proposition 4.11).

Lemma 4.9 Under the assumption that $\Theta_1 \geq \Theta_0$, both $p_1(k)/p_0(k)$ and $p_2(k)/p_0(k)$ are increasing in k, for $k \leq N$.

Proof.

From (4.1) and (4.2), we have

$$\frac{p_1(k)}{p_0(k)} = \frac{E[\Theta_1^k(1-\Theta_1)^{N-k}]}{E[\Theta_0^k(1-\Theta_0)^{N-k}]}. \qquad (4.22)$$

Because $\Theta_0 \leq \Theta_1$ as assumed, we have

$$\Theta_0^k(1-\Theta_0)^{N-k}\Theta_1^{k-1}(1-\Theta_1)^{N-k+1}$$
$$\leq \Theta_0^{k-1}(1-\Theta_0)^{N-k+1}\Theta_1^k(1-\Theta_1)^{N-k}.$$

Taking expectations on both sides and taking into account the independence between Θ_0 and Θ_1, we have shown the increasingness of the right-hand side of (4.22) and hence the increasingness of $p_1(k)/p_0(k)$ in k.

Next, we show that $\bar{p}_2(k)/\bar{p}_0(k)$ is increasing in k. Consider any realization of Θ_0 and Θ_1, i.e., $\Theta_0 = \theta_0 \leq \Theta_1 = \theta_1$. Clearly, $\bar{p}_0(k)$ is the probability mass function (evaluated at k) of a binomial random variable, denoted B_0, which can be expressed as follows:

$$B_0 = I_1(\theta_0) + \cdots + I_N(\theta_0),$$

where $I_k(\theta)$ denotes a binary (0-1) random variable that equals 1 w.p. θ, and the I_ks are independent. On the other hand, $\bar{p}_2(k)$ is the probability mass function (also evaluated at k) of another random variable, denoted B_2, which is equal to

$$B_2(j) := I_1(\theta_0) + \cdots + I_{j-1}(\theta_0) + I_j(\theta_1) + \cdots + I_N(\theta_1)$$

with probability

$$[\exp(-\frac{j-1}{N}\lambda) - \exp(-\frac{j}{N}\lambda)]/[1 - \exp(-\lambda)]$$

for $j = 1, ..., N$. (Note here that we condition on the state shift taking place while processing the jth component.) Direct verification establishes that $I(\theta_0) \leq_{\mathrm{lr}} I(\theta_1)$ (Definition 2.1), and that this likelihood ratio ordering is preserved under convolution of the independent binary random variables in question (refer to [49, 85]). Consequently, we have $B_0 \leq_{\mathrm{lr}} B_2(j)$ for all $j = 1, ..., n$; and hence, $B_0 \leq_{\mathrm{lr}} B_2$. That is,

$$\bar{p}_0(k)\bar{p}_2(k-1) \leq \bar{p}_0(k-1)\bar{p}_2(k),$$

for all k, which is the desired monotonicity. $\square$

Lemma 4.10 $Y_i = f_1(Y_{i-1}, D_i)$ following Lemma 4.1 satisfies the following properties:
(i) $f_1(\cdot, \cdot)$ is an increasing function in both components;
(ii) Y_i is stochastically increasing in Y_{i-1}; and
(iii) $\mathsf{E}[Y_i \mid Y_{i-1}, a_{i-1} = 0, b_{i-1} = 0] \geq Y_{i-1}$.

Proof. (i) From (4.5), we have

$$\begin{aligned}
&f_1(Y_{i-1}, \mathcal{D}_i) \\
&= 1 - \frac{(1-Y_{i-1})p_0(\mathcal{D}_i)e^{-\lambda}}{Y_{i-1}p_1(\mathcal{D}_i) + (1-Y_{i-1})[p_2(\mathcal{D}_i)(1-e^{-\lambda}) + p_0(\mathcal{D}_i)e^{-\lambda}]} \\
&= 1 - \left[\frac{Y_{i-1}}{1-Y_{i-1}}\frac{p_1(\mathcal{D}_i)}{p_0(\mathcal{D}_i)}e^{\lambda} + \frac{p_2(\mathcal{D}_i)}{p_0(\mathcal{D}_i)}(e^{\lambda}-1) + 1\right]^{-1}.
\end{aligned}$$

Because $Y_{i-1}/(1 - Y_{i-1})$ is increasing in Y_{i-1}, $f_1(\cdot, \cdot)$ is increasing in its first component. From the preceding expression, it is also obvious that the increasing property of $f_1(\cdot, \cdot)$ with respect to its second component is guaranteed if the ratios, $p_1(\mathcal{D}_i)/p_0(\mathcal{D}_i)$ and $p_2(\mathcal{D}_i)/p_0(\mathcal{D}_i)$, are increasing in $\mathcal{D}_i$. But this follows from Lemma 4.9.

(ii) Suppose $a_{i-1} = 0$. Because D_i also depends on Y_{i-1}, from Lemma 4.1 we can write $Y_i = f_1(Y_{i-1}, D_i(Y_{i-1}))$. Note the following:

$$
\begin{aligned}
&P[D_i(y) = k] \\
= &\; y p_1(k) + (1 - y)e^{-\lambda} p_0(k) + (1 - y)(1 - e^{-\lambda})p_2(k) \\
= &\; y\{p_1(k) - p_2(k) + e^{-\lambda}[p_2(k) - p_0(k)]\} \\
&\; + e^{-\lambda} p_0(k) + (1 - e^{-\lambda})p_2(k).
\end{aligned}
\tag{4.23}
$$

Clearly, $D_i(y)$ is stochastically increasing in y, taking into account the following inequalities:

$$
\sum_{k=\ell}^{N} p_1(k) \geq \sum_{k=\ell}^{N} p_0(k) \quad \text{and} \quad \sum_{k=\ell}^{N} p_2(k) \geq \sum_{k=\ell}^{N} p_0(k)
$$

for all $\ell = 0, 1, ..., N$. That is, the random variable associated with the probability mass function $\{p_0(k)\}$, is stochastically smaller than both random variables associated with $\{p_1(k)\}$ and $\{p_2(k)\}$—a conclusion that follows from the proof of Lemma 4.9, where the stronger likelihood ratio ordering was established. Because $f_1(\cdot, \cdot)$ is increasing in both components, it follows that Y_i is stochastically increasing in Y_{i-1}. When $a_{i-1} = 1$, we have $Y_i = f(0, D_i(Y_{i-1}))$, and the same argument applies.

(iii) It is easy to see that

$$
\begin{aligned}
&E[Y_i \mid Y_{i-1}, a_{i-1} = 0, b_{i-1} = 0] \\
= &\; f_0(Y_{i-1}) \\
= &\; Y_{i-1} + (1 - Y_{i-1})(1 - e^{-\lambda}) \\
\geq &\; Y_{i-1}. \quad \square
\end{aligned}
\tag{4.24}
$$

Proposition 4.11 Both $V_\alpha(1, y)$ and $V_\alpha(2, y)$, defined through (4.16) and (4.17), are increasing and concave in y.

Proof. From Lemma 4.7, along with (4.18) and (4.19), we know that V_α is the limit of recursions governed by the operators min and plus, both preserving increasing concavity. (Refer to the SSICV property following Definition 2.10.) Hence, what remains is to argue, inductively from the recursions, that [in (4.19)] both $V_\alpha^{m-1}(1, f_0(y))$ and $E[V_\alpha^{m-1}(1, f_1(y, D))]$ are increasing and concave in y, given the increasing concavity of $V_\alpha^{m-1}(1, \cdot)$ with respect to its second argument.

The increasing concavity of $V_\alpha^{m-1}(1, f_0(y))$ in y is obvious, because $f_0(y)$ is increasing and linear in y [cf. (4.4)]. To show the increasing concavity of

$E[V_\alpha^{m-1}(1, f_1(y, D))]$ in y amounts to proving the increasing concavity of $\phi(y) := E[V(f_1(y, D(y)))]$, given the increasing concavity of V, where, to lighten notation, we write $V(\cdot) := V_\alpha^{m-1}(1, \cdot)$. Also note that D depends on y as well, because the number of defectives in a batch depends on the machine state at the beginning of the period.

The increasing property is immediate, because V is increasing, and so is f_1 (with respect to both arguments, cf. Lemma 4.10 (i)). As pointed out in the proof of Lemma 4.10, $D(y)$ is stochastically increasing in y. Hence, $\phi(y)$ is increasing in y.

To show concavity, write

$$\phi(y) = \sum_{k=1}^{n} V(f_1(y, k))q(y, k),$$

where

$$
\begin{aligned}
q(y, k) &:= P[D_i(y) = k] \\
&= yp_1(k) + (1 - y)[p_2(k)(1 - e^{-\lambda}) + p_0(k)e^{-\lambda}].
\end{aligned}
\tag{4.25}
$$

Taking (second) derivatives with respect to y, and omitting the arguments of V, f_1 and q for simplicity, we have

$$
\begin{aligned}
&\phi''(y) \\
&= \sum_k [V''(f_1')^2 q + V'f_1''q + 2V'f_1'q' + Vq''] \\
&\leq \sum_k [V'f_1''q + 2V'f_1'q'],
\end{aligned}
\tag{4.26}
$$

because $V'' \leq 0$ (concavity) and $q'' = 0$ (linear in y). Furthermore, from (4.4) and (4.5), we have

$$f_1(y, k)q(y, k) = [p_1(k) - p_2(k)]y + p_2(k)f_0(y).\tag{4.27}$$

Again, taking derivatives with respect to y on both sides and omitting the arguments, we have

$$f_1''q = -2f_1'q' - f_1q'' + p_2f_0'' = -2f_1'q',$$

because both q and f_0 are linear in y. Substituting this into (4.26), we have

$$\phi''(y) \leq \sum_k [V'f_1''q + 2V'f_1'q'] = 0$$

and hence the desired concavity. $\square$

4.3.3 Optimal Policies

We start with the optimal policy for machine revision.

Theorem 4.12 There exists a threshold-type optimal policy for the machine revision decision. Specifically, it is optimal to revise the machine in state y if and only if $y \geq y_\alpha^R$, where

$$y_\alpha^R := \sup \{y \leq 1 : V_\alpha(2, y) \leq C_R + V_\alpha(2, 0)\}.$$

Proof. From Theorem 4.8, it is optimal to revise the machine in state y if and only if $C_R + V_\alpha(2, 0) \leq V_\alpha(2, y)$. Because $V_\alpha(2, y)$ is increasing in y (Proposition 4.11), the inequality holds if and only if $y \geq y_\alpha^R$. □

The structure of the optimal inspection policy is more complicated. It is known that even in special cases of our model (e.g., [95]) the optimal inspection policy cannot be characterized by a single threshold. We start with a lemma.

Lemma 4.13 For $j = 1, 2$, we have $\mathsf{E}[V_\alpha(j, f_1(y, D))] \leq V_\alpha(j, f_0(y))$.

Proof. Notice that $\mathsf{E}f_1(y, D) = f_0(y)$, the desired result follows from Jensen's inequality, because $V_\alpha(j, \cdot)$ is concave in its second argument, for $j = 1, 2$, following Proposition 4.11. □

Define
$$y_1^I := \sup \{y \leq y_\alpha^R : r(y, 0) - r(y, 1) \leq C_I\} \tag{4.28}$$

and

$$\begin{aligned} y_2^I := \ & \sup \{y \leq 1 : r(y, 1) + \alpha \mathsf{E}[V_\alpha(f_1(y, D))] \\ & \leq C_R + r(0, 0) - C_I + \alpha V_\alpha(f_0(0))\}. \end{aligned} \tag{4.29}$$

The following theorem describes a *partial* threshold structure of the optimal inspection policy.

Theorem 4.14 (i) Suppose

$$C_I + r(0, 1) + \alpha \mathsf{E}[V_\alpha(f_1(0, D))] \leq r(0, 0) + \alpha V_\alpha(f_0(0)). \tag{4.30}$$

Then it is optimal to do inspection in any state y that satisfies $y > y_1^I$.
(ii) Suppose

$$C_I + r(0, 1) + \alpha \mathsf{E}[V_\alpha(f_1(0, D))] > r(0, 0) + \alpha V_\alpha(f_0(0)). \tag{4.31}$$

(ii-a) If $y_2^I > y_1^I$, then it is optimal to do inspection (but no revision) in any state $y \in (y_1^I, y_2^I]$ and not to do inspection (do revision instead) for $y \in (y_2^I, 1]$.
(ii-b) If $y_2^I \leq y_1^I$, then it is optimal not to do inspection for $y \in (y_2^I, 1]$.

(iii) In the special case of $C_I \leq r(0,0) - r(0,1)$, it is optimal to inspect every batch (i.e., to do inspection in any state y).

Proof. The key is to examine the optimality equation in (4.14) which is repeated here:

$$
\begin{aligned}
V_\alpha(y) = \min\{ & C_R + C_I + r(0,1) + \alpha E[V_\alpha(f_1(0,D))], \\
& C_R + r(0,0) + \alpha V_\alpha(f_0(0)), \\
& C_I + r(y,1) + \alpha E[V_\alpha(f_1(y,D))], \\
& r(y,0) + \alpha V_\alpha(f_0(y))\}.
\end{aligned}
\tag{4.32}
$$

Notice that

$$
E[V_\alpha(f_1(y,D))] \leq V_\alpha(f_0(y)) \qquad \text{for any } y,
$$

under condition in (i), the second term under the min operator in (4.32) dominates the first one. Also, when $y > y_1^I$, either $r(y,0) > C_I + r(y,1)$, and hence the fourth term dominates the third one; or $y > y_1^I = y_\alpha^R$. In the second case, from the definition of y_α^R we have

$$
\begin{aligned}
& \min\{C_I + r(y,1) + \alpha E[V_\alpha(1, f_1(y,D))], r(y,0) + \alpha V_\alpha(1, f_0(y))\} \\
> \ & C_R + \min\{C_I + r(0,1) + \alpha E[V_\alpha(1, f_1(0,D))], r(0,0) + \alpha V_\alpha(1, f_0(0))\} \\
= \ & C_R + C_I + r(0,1) + \alpha E[V_\alpha(1, f_1(0,D))].
\end{aligned}
$$

Hence, in any event the minimum is between the first and third terms, both of which involve inspection.

Similarly, for (ii-a), when $y \in (y_1^I, y_2^I]$, the minimum in (4.32) is attained by the third term; hence it is optimal to do inspection (but no revision). On the other hand, when $y \in (y_2^I, 1]$, the minimum is reached at the second term; hence it is optimal not to do inspection (but to do revision instead).

For (ii-b), when $y > y_2^I$, the minimum in (4.32) is between the second and fourth terms (in particular, the third term dominates the second term), and neither involves inspection.

Finally, (iii) is a special case of (i), with the fourth term dominating the third for any y. $\square$

From the revision and inspection policies stated in Theorem 4.14, we have the following.

Corollary 4.15 In all cases of Theorem 4.14, the revision threshold always dominates the inspection threshold. In particular, $y_\alpha^R = y_2^I$ in (ii-a). $\square$

The cases that are left out in Theorem 4.14 are $y \leq y_1^I$ in (i) and (ii-a), and $y \leq y_2^I \leq y_1^I$ in (ii-b). In these cases, the minimum in (4.32) involves both the third and fourth terms, which can cross over each other many times in general, resulting in many thresholds. More specifically, the difficulty is that the difference,

$$
V_\alpha(f_0(y)) - E[V_\alpha(f_1(y,D))],
\tag{4.33}
$$

is not necessarily concave (or convex), although we do know that each term is concave in y.

Now, suppose the difference in (4.33) is itself concave in y. Then, because $r(y, 0)$ and $r(y, 1)$ are all linear in y, the following expression,

$$r(y, 0) - r(y, 1) + \alpha V_\alpha(f_0(y)) - \alpha \mathsf{E}[V_\alpha(f_1(y, D))],$$

is concave in y, and hence it can only cross over the constant C_I at most two times: first from below and then from above. Therefore, under this concavity condition, there will be at most two crossover points for the third and fourth terms in (4.32), which we denote as $y_a^I \le y_b^I$. Hence, the third term is smaller than the fourth term when $y \in [y_a^I, y_b^I]$, which favors inspection; while the fourth term is smaller when $y \in [0, y_a^I)$ or $y \in (y_b^I, 1]$.

Therefore, when the difference in (4.33) is concave in y, case (ii-a) in Theorem 4.14 takes the following form (taking into account Corollary 4.15): $y_2^I = y_b^I = y_\alpha^R$, $y_1^I \in [y_a^I, y_b^I)$. Hence it is optimal to do revision but no inspection for $y \in (y_b^I, 1]$, do inspection but no revision for $y \in [y_a^I, y_b^I]$, and do neither for $y \in [0, y_a^I)$.

For case (ii-b), we have: $y_1^I \ge y_2^I \ge y_b^I$; depending on the value of y_α^R, we have the following cases:

(ii-b-1) $y_\alpha^R \le y_a^I$: do nothing in $[0, y_\alpha^R]$, and do revision only in $(y_\alpha^R, 1]$;

(ii-b-2) $y_a^I < y_\alpha^R \le y_b^I$: do nothing in $[0, y_a^I)$, do inspection only in $[y_a^I, y_\alpha^R]$, and do revision only in $(y_\alpha^R, 1]$; and

(ii-b-3) $y_b^I < y_\alpha^R \le 1$: do nothing in $[0, y_a^I)$, do inspection only in $[y_a^I, y_b^I]$, do nothing in $(y_b^I, y_\alpha^R]$, and do revision only in $(y_\alpha^R, 1]$.

For case (i) of Theorem 4.14, note that there is only one crossover point y_b^I, because the given condition indicates that the fourth term under the minimum in (4.32) already dominates the third term at $y = 0$. (Also note that in this case the second term dominates the first term, which corresponds to doing both inspection and revision.) Hence, the optimal actions are: if $y_b^I < y_\alpha^R$, then do inspection in $[0, y_b^I)$, do nothing in $[y_b^I, y_\alpha^R]$ and do both inspection and revision in $(y_\alpha^R, 1]$. If $y_b^I \ge y_\alpha^R$, then do inspection in $[0, y_\alpha^R)$, and do both inspection and revision in $(y_\alpha^R, 1]$. The policies under different cases are summarized in Figures 4.2 and 4.3.

To illustrate the special cases, one (rather trivial) example is $\Theta_0 = \Theta_1$, which leads to $p_0(k) = p_1(k) = p_2(k)$ for all k. Hence $f_1(y, k) = f_0(y)$ for all k, and $V_\alpha(f_0(y)) = \mathsf{E}[V_\alpha(f_1(y, D))]$. That is, the difference in (4.33) is zero. A more interesting example is either

$$p_0(0) = 1 \quad \text{and} \quad p_1(0) = p_2(0) = 0 \tag{4.34}$$

(i.e., no defective in a perfect state, and at least one defective otherwise); or

$$p_0(0) = p_2(0) = 1 \quad \text{and} \quad p_1(0) = 0 \tag{4.35}$$

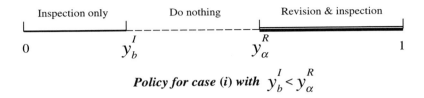

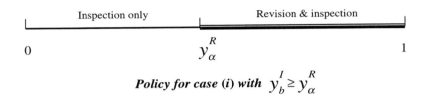

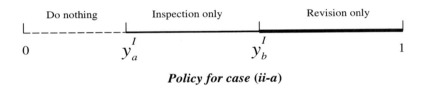

FIGURE 4.2. Policy for Case (i) and Case (ii-a).

(same as earlier, except that state shift will not affect the quality of the product; refer to Remark 4.16). In both cases, for each k, $f_1(y,k) := c(k)$ is a constant; hence,

$$E[V_\alpha(f_1(y,D))] = \sum_{k=1}^{N} V_\alpha(f_1(y,k))q(y,k) = \sum_{k=1}^{N} V_\alpha(c(k))q(y,k)$$

is linear in y, because $q(y,k)$ is linear in y for each k (refer to (4.25)), and hence the difference in (4.33) is concave.

Remark 4.16 In Ross [75], the key assumptions are as follows: $n = 1$, $\Theta_0 = 0$, and $\Theta_1 = 1$, and in particular state shift within a period will not affect the quality of the product (hence, it is defective if and only if the machine state at the beginning of the period is out of control).

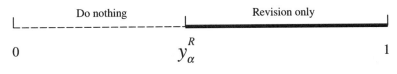

Policy for case (ii-b-1)

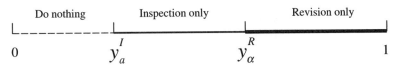

Policy for case (ii-b-2)

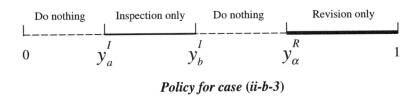

Policy for case (ii-b-3)

FIGURE 4.3. Policy for Case (ii-b-1) to Case (ii-b-3).

It is easy to verify that in this case $r(0,0) = r(0,1) = 0$, (4.35) holds, and hence

$$\mathsf{E}[V_\alpha(f_1(0,D))] = V_\alpha(f_1(0,0)) = V_\alpha(f_0(0)).$$

Therefore, the optimality equation in (4.14) simplifies to the following:

$$\begin{aligned}
V_\alpha(y) = \min\{ &C_R + \alpha V_\alpha(f_0(0)), \\
&C_I + r(y,1) + \alpha\mathsf{E}[V_\alpha(f_1(y,D))], \\
&r(y,0) + \alpha V_\alpha(f_0(y))\}.
\end{aligned}$$

This reduces to the special case (ii) of Theorem 4.14 discussed earlier.

Indeed, the optimal policy in [75] is characterized in general by four intervals as illustrated in (ii-b-3), with the four intervals corresponding, respectively, to the actions: do nothing, do inspection only, do nothing, and do revision only.

Finally, note that if the difference in (4.33) is convex in y, the earlier discussion still applies (with the inspection and no-inspection intervals switching places).

4.4   Average-Cost Model

For the average-cost model, we have the following.

Theorem 4.17 (i) There exists a bounded function $h(y)$ and a constant g satisfying

$$
\begin{aligned}
g + h(y) \\
= \quad \min\{ &C_R + C_I + r(0,1) + \mathsf{E}[h(f_1(0,D))], \\
&C_R + r(0,0) + h(f_0(0)), \\
&C_I + r(y,1) + \mathsf{E}[h(f_1(y,D))], \\
&r(y,0) + h(f_0(y))\}.
\end{aligned} \tag{4.36}
$$

The optimal long-run average-cost, starting from state y, is $\bar{V}(y) = g$. Furthermore, any stationary policy $\bar{\pi}$ that prescribes actions minimizing the right-hand side of equation (4.36) is optimal.
(ii) There exists an increasing sequence $\alpha_n \to 1$ such that

$$
h(y) = \lim_{n \to \infty} [V_{\alpha_n}(y) - V_{\alpha_n}(0)] \quad \text{and} \quad \lim_{\alpha \to 1} (1-\alpha) V_\alpha(0) = g.
$$

Proof. From Lemma 4.3 and 4.4, it suffices to prove that there exists a constant K such that $|V_\alpha(y) - V_\alpha(0)| \le K$ for any $\alpha \in (1,0)$. Because

$$
V_\alpha(y) = V_\alpha(1,y)
$$

and from (4.16),

$$
V_\alpha(1,y) \le C_R + V_\alpha(2,0),
$$

and

$$
V_\alpha(1,0) = V_\alpha(2,0),
$$

we have

$$
\begin{aligned}
0 &\le V_\alpha(y) - V_\alpha(0) = V_\alpha(1,y) - V_\alpha(1,0) \\
&\le C_R + V_\alpha(2,0) - V_\alpha(2,0) = C_R.
\end{aligned}
$$

Hence, we can let $K = C_R$. $\square$

Let $H(1,y) = h(y)$, and define $H(2,y)$ through

$$
\begin{aligned}
g + H(2,y) \\
= \quad \min\{ &C_I + r(y,1) + \mathsf{E}[H(1, f_1(y,D))], \\
&r(y,0) + H(1, f_0(y))\}.
\end{aligned}
$$

Then, similar to the discussion in §4.3.1, we have

$$g + H(1, y) = \min\{C_R + H(2, 0), H(2, y)\}.$$

As in the discounted-cost model, we define a policy $\bar{\pi}$ as follows: in period i, if the state is y, choose $a_i = 1$ if and only if

$$C_R + H(2, 0) \leq H(2, y); \tag{4.37}$$

furthermore, if $a_i = 1$ is chosen, then choose $b_i = 1$ if and only if

$$C_I + r(0, 1) + \mathsf{E}[H(1, f_1(0, D))] \leq r(0, 0) + H(1, f_0(0)), \tag{4.38}$$

and if $a_i = 0$ is chosen, then choose $b_i = 1$ if and only if

$$C_I + r(y, 1) + \mathsf{E}[H(1, f_1(y, D))] \leq r(y, 0) + H(1, f_0(y)). \tag{4.39}$$

Following exactly the same argument as in Theorem 4.8, we have the following.

Theorem 4.18 The policy $\bar{\pi}$ prescribes actions that minimize the right-hand side of the optimality equation (4.36) and therefore is a stationary optimal policy for the average-cost problem. □

Following Theorem 4.17 (ii), making use of Proposition 4.11, and following the proof of Lemma 4.13, we can show that, for $j = 1, 2$, $H(j, y)$ is increasing and

$$\mathsf{E}[H(j, f_1(y, D)) \,|\, Y_0 = y] \leq H(j, f_0(y)).$$

These, along with Theorem 4.18 and the inequalities in (4.37-4.39), yield the following structure of the optimal policy for the average-cost model.

Theorem 4.19 (i) For machine revision, it is optimal to revise the machine in state y if and only if $y \geq \bar{y}^R$, where

$$\bar{y}^R := \sup \{y \leq 1 : H(2, y) \leq C_R + H(2, 0)\}.$$

(ii) The optimal inspection policy has the same structure as described in Theorem 4.14; in particular, it follows all the thresholds specified there. □

The discussion following Theorem 4.14 obviously applies here as well.

4.5   A Special Case: No Inspection Cost

Suppose $C_I = 0$, i.e., inspection incurs no cost. Then, from Lemma 4.13, it is optimal to inspect every batch of products. Hence, we only have to make the revision decision in each period. Letting $b_i \equiv 1$, we can apply the results in the previous sections. For convenience, denote $r(y, 1)$ as $r(y)$. From Lemma 4.6 and Theorems 4.12, 4.17, and 4.19, we have the following.

Theorem 4.20 (i) For the discounted-cost model, the optimal policy is to revise the machine if and only if $y \geq y_\alpha$, where

$$y_\alpha := \sup\{y \leq 1 : r(y) + \alpha \mathsf{E}[V_\alpha(f_1(y, D))] \leq C_R + V_\alpha(0)\}. \qquad (4.40)$$

(ii) For the average-cost model, it is optimal to revise the machine at y if and only if $y \geq \bar{y}$, with

$$\bar{y} := \sup\{y \leq 1 : r(y) + \mathsf{E}[h(f_1(y, D))] \leq C_R + r(0) + \mathsf{E}[h(f_1(0, D))]\}. \qquad \square$$

Furthermore, we can relate the two models by establishing that the optimal threshold in (4.40) is decreasing in the discount factor α and converges to $\bar{y}$ as $\alpha \to 1$. We need a lemma.

Lemma 4.21 For $1 > \alpha' > \alpha > 0$, we have
(i) $(1 - \alpha)V_\alpha(0) \leq (1 - \alpha')V_{\alpha'}(0)$; and
(ii) $(1 - \alpha)[V_\alpha(0) + C_R] \geq (1 - \alpha')[V_{\alpha'}(0) + C_R]$.

Proof. (i) Consider a variation of the original problem. Suppose, in addition to the usual machine revision, which requires a cost C_R, we now have cost free opportunities, independent of the system state, to revise the machine at the beginning of *some* periods and the time between two consecutive opportunities is a geometric random variable with parameter α, denoted T_α. That is, $\mathsf{P}[T_\alpha = j] = (1 - \alpha)\alpha^{j-1}$ for $j = 1, 2, \ldots$. Suppose we want to minimize the long-run average-cost for this problem. Because $r(y)$, the cost related to defects, is increasing in y, it is easy to see that the more cost free opportunities we have, the less cost we will incur, and an optimal policy will make use of all these free opportunities to revise the machine. Suppose π is a policy that uses all of these opportunities. For convenience, let $R(Y_i, a_i)$ denote the cost at time period i, when the state is Y_i and control action a_i is chosen. From the standard theory of renewal processes (e.g., Ross [73]), the long-run average cost under the stationary policy π, denoted as $\bar{V}_{\pi,\alpha}(0)$, can be obtained as follows:

$$\bar{V}_{\pi,\alpha}(0) = \mathsf{E}_\pi\left[\sum_{j=1}^{T_\alpha} R(Y_{j-1}, a_{j-1}) \,|\, Y_0 = 0\right]/\mathsf{E}[T_\alpha].$$

Because

$$\mathsf{E}_\pi\left[\sum_{j=1}^{T_\alpha} R(Y_{j-1}, a_{j-1}) \,|\, Y_0 = 0\right]$$

$$= \sum_{j=1}^{\infty} \mathsf{E}_\pi[R(Y_{j-1}, a_{j-1}) \,|\, Y_0 = 0]\mathsf{P}[T_\alpha \geq j]$$

$$= \sum_{j=1}^{\infty} \mathsf{E}_\pi[R(Y_{j-1}, a_{j-1}) \,|\, Y_0 = 0]\alpha^{j-1}$$

$$= V_\alpha(\pi, 0)$$

(T_α being independent of the states) and $E[T_\alpha] = (1 - \alpha)^{-1}$, we have $\bar{V}_{\pi,\alpha}(0) = (1 - \alpha)V_\alpha(\pi, 0)$. Therefore, for a given α, the optimal average-cost will be

$$\bar{V}_\alpha(0) := \inf_\pi \bar{V}_{\pi,\alpha}(0) = (1 - \alpha)V_\alpha(0).$$

On the other hand, from the definition of T_α, for larger α, we will have less opportunity to revise the machine for free and hence a larger average-cost, i.e., $\bar{V}_{\pi,\alpha}(0) \leq \bar{V}_{\pi,\alpha'}(0)$ for $\alpha' \geq \alpha$. This implies the inequality in (i).

(ii) Consider another variation of the original problem. Assume that, in lieu of free revising opportunities, there are now *shocks* that will drive the machine out of control and force us to revise it with a cost C_R. Suppose the interarrival time between two consecutive shocks is T_α, which as before is a geometric random variable with parameter α, independent of the system states. Denote the long-run average-cost under the policy π as $\hat{V}_{\pi,\alpha}(0)$. Note that each time a shock occurs, we pay an extra cost C_R; hence, under policy π, the cost between two shocks is

$$E_\pi[\sum_{j=1}^{T_\alpha} R(Y_{j-1}, a_{j-1}) \mid Y_0 = 0] + C_R = V_\alpha(\pi, 0) + C_R$$

and

$$\hat{V}_{\pi,\alpha}(0) = (1 - \alpha)[V_\alpha(\pi, 0) + C_R].$$

Observe that, for a larger α, we have fewer shocks and hence a smaller average-cost, i.e., $\hat{V}_{\pi,\alpha}(0) \geq \hat{V}_{\pi,\alpha'}(0)$ for any policy π. This implies the inequality in (ii). $\square$

Proposition 4.22 The optimal threshold y_α is decreasing in the discount factor α. That is, for $1 > \alpha' > \alpha > 0$, $y_{\alpha'} \leq y_\alpha$.

Proof. By contradiction. Suppose $y_{\alpha'} > y_\alpha$, and π' is the optimal threshold control policy corresponding to α'. Consider any state y with $y_{\alpha'} > y \geq y_\alpha$. Let M be the first time the system enters some state y' with $y' \geq y_{\alpha'}$, starting from y and under policy π'. We have

$$C_R + V_\alpha(0) \leq E_{\pi'}[\sum_{j=0}^{M-1} \alpha^j r(Y_j) + \alpha^M C_R + \alpha^M V_\alpha(0) \mid Y_0 = y]$$

and

$$C_R + V_{\alpha'}(0) > E_{\pi'}[\sum_{j=0}^{M-1} \alpha'^j r(Y_j) + \alpha'^M C_R + \alpha'^M V_{\alpha'}(0) \mid Y_0 = y].$$

The first inequality follows from the fact that (under the discount factor α) to make a revision now is better than to make one at some time in the future, because $y \geq y_\alpha$. The second inequality follows from the fact that under the discount factor α', it is optimal not to make the revision until M, because $y < y'_\alpha$. The two inequalities can be simplified to:

$$(1 - \mathsf{E}_{\pi'}[\alpha^M \mid Y_0])[C_R + V_\alpha(0)] \leq \mathsf{E}_{\pi'}[\sum_{j=0}^{M-1} \alpha^j r(Y_j) \mid Y_0]$$

and

$$(1 - \mathsf{E}_{\pi'}[\alpha'^M \mid Y_0])[C_R + V_{\alpha'}(0)] > \mathsf{E}_{\pi'}[\sum_{j=0}^{M-1} \alpha'^j r(Y_j) \mid Y_0].$$

Combining these two, we have

$$(1 - \mathsf{E}_{\pi'}[\alpha'^M \mid Y_0])[C_R + V_{\alpha'}(0)] - (1 - \mathsf{E}_{\pi'}[\alpha^M \mid Y_0])[C_R + V_\alpha(0)]$$
$$> \mathsf{E}_{\pi'}[\sum_{j=0}^{M-1} (\alpha'^j - \alpha^j) r(Y_j) \mid Y_0]. \tag{4.41}$$

We will show that for any given $M = m$, inequality (4.41) *cannot* hold. Denote the left-hand side of (4.41) by LHS. For any given m, we have

$$
\begin{aligned}
LHS &= \frac{(1 - \alpha'^m)}{1 - \alpha'}(1 - \alpha')[V_{\alpha'}(0) + C_R] - \frac{(1 - \alpha^m)}{1 - \alpha}(1 - \alpha)[V_\alpha(0) + C_R] \\
&\leq [\frac{(1 - \alpha'^m)}{1 - \alpha'} - \frac{(1 - \alpha^m)}{1 - \alpha}](1 - \alpha)[V_\alpha(0) + C_R] \\
&= \sum_{j=0}^{m-1} (\alpha'^j - \alpha^j)(1 - \alpha)[V_\alpha(0) + C_R], \tag{4.42}
\end{aligned}
$$

where the inequality is from Lemma 4.21 (ii). Because under the discount factor α, when $y \geq y_\alpha$, it is better to revise the machine immediately than to do so one period later, we have

$$C_R + V_\alpha(0) \leq r(y) + \alpha[C_R + V_\alpha(0)].$$

From Lemma 4.5, we know $r(y)$ is a linear increasing function of y. Hence, applying Lemma 4.10 (iii) repeatedly, we have

$$\mathsf{E}_{\pi'}[r(Y_j) \mid Y_0 = y] \geq r(y).$$

Substituting this into the last inequality, we have

$$(1 - \alpha)[C_R + V_\alpha(0)] \leq \mathsf{E}_{\pi'}[r(Y_j) \mid Y_0 = y]$$

for any $1 \leq j \leq m$. This implies, along with (4.42), that for any m,

$$(1 - \alpha'^m)[C_R + V_{\alpha'}(0)] - (1 - \alpha^m)[C_R + V_\alpha(0)]$$
$$\leq E_{\pi'}[\sum_{j=0}^{m-1}(\alpha'^j - \alpha^j)r(Y_j) \,|\, Y_0 = y].$$

This contradicts inequality (4.41). Therefore, under the discount factor α', it is also optimal to revise the machine in state y, and hence $y_{\alpha'} \leq y_\alpha$. $\square$

The intuition behind this result is this: for a larger discount factor, the future cost related to defectives becomes more significant, so we are more willing to pay a cost to revise the machine to reduce the future cost of defects.

Theorem 4.23 $y_\alpha \to \bar{y}$ as $\alpha \to 1$.

Proof. By Theorem 4.17 (ii), there exists an increasing sequence $\{\alpha_n\} \to 1$ such that

$$h(y) = \lim_{n \to \infty}[V_{\alpha_n}(y) - V_{\alpha_n}(0)].$$

From (4.40), noting that $V_\alpha(0) = r(0) + \alpha E[V_\alpha(f_1(0, D))]$, we have

$$y_{\alpha_n} = \sup\{y \leq 1 : C_R + r(0) + \alpha_n[EV_{\alpha_n}(f_1(0, D)) - V_{\alpha_n}(0)]$$
$$\geq r(y) + \alpha_n[EV_{\alpha_n}(f_1(y, D)) - V_{\alpha_n}(0)]\}.$$

Because the optimal policy is of threshold type,

$$\{y : C_R + r(0) + \alpha_n[EV_{\alpha_n}(f_1(0, D)) - V_{\alpha_n}(0)]$$
$$\leq r(y) + \alpha_n[EV_{\alpha_n}(f_1(y, D)) - V_{\alpha_n}(0)]\}$$
$$= [y_{\alpha_n}, 1].$$

Note that y_{α_n} is decreasing as $\alpha_n \to 1$, and that for a given y, the number of possible y', the states at the beginning of the next period, is finite. Hence, we can exchange the limit and the expectation:

$$\lim_{n \to \infty} y_{\alpha_n}$$
$$= \lim_{n \to \infty} \sup\{y \leq 1 : C_R + r(0) + \alpha_n[EV_{\alpha_n}(f_1(0, D)) - V_{\alpha_n}(0)]$$
$$\geq r(y) + \alpha_n[EV_{\alpha_n}(f_1(y, D)) - V_{\alpha_n}(0)]\}$$
$$= \sup\{y \leq 1 : C_R + r(0) + E[h(f_1(0, D))]$$
$$\geq r(y) + E[h(f_1(y, D))]\}$$
$$= \bar{y}.$$

Because y_α is decreasing in α, $\lim_{\alpha \to 1} y_\alpha = \lim_{n \to \infty} y_{\alpha_n} = \bar{y}$. $\square$

4.6 Notes

A traditional approach to process control is the control chart technique: an in control area is specified via an upper limit and a lower limit, in terms of certain specific process measurements. Each time a measurement is taken, it is charted. If it falls within the in control area, the machine is deemed in control and the process is allowed to continue; otherwise the process is halted and a revision takes place. The centerpiece of this technique is obviously the control limits, along with the process measurement. However, despite a large body of literature on this subject (refer to, e.g., Shewhart [86], Thompson and Koronacki [97], and the numerous references there), there does not appear to be any systematic approach to determining both the process measurements and the control limits that exploit all the relevant information, in particular the dynamics of the process, to achieve optimality in terms of minimizing the overall costs of inspection and revision.

A few earlier studies in the literature that take a dynamical approach similar to what is described in this chapter do so under more restrictive assumptions. For example, Ross [75] considered single-unit production $(N = 1)$ and established an optimal policy characterized by four intervals (of machine-state probabilities) under the additional assumption that the defective rate is deterministic and binary; the item produced in each period is defective if and only if the machine is out of control at the beginning of the period. (Refer to more details in Remark 4.16.) Other studies also considered single-unit production and ignored inspection costs, e.g., Girshick and Rubin [37] and Taylor [95].

The model presented here is based on Yao and Zheng [108]. In allowing batch size (i.e., $N \geq 1$), the model can accommodate more interesting applications (in batch manufacturing, for example). The batch size also significantly changes the problem structure (the machine state can shift within the period, resulting in two product groups with different defective rates) and makes the treatment of the defective rates as random variables more challenging. (The case of $N = 1$ effectively reduces random defective rates to their means and is hence no different from the case with deterministic defective rates.) We have successfully studied both issues. In addition, we have developed several new results. In particular, in the presence of inspection cost, we have shown that the optimal policy for machine revision is still of threshold type, while the optimal policy for inspection has a more involved structure characterized by several threshold values.

If the inspection cost is charged for *each* inspected item in the batch, then the problem becomes more complex: the inspection itself becomes an optimal stopping problem, which intertwines with the machine revision problem. For example, in general we may choose to inspect only *part* of the batch to improve quality as well as to update our knowledge about the machine state. One crucial but difficult issue is to find an optimality equation like the one in (4.14). On the other hand, in the case of *pure* inspection, i.e., without machine revision, under a very general cost structure for defective products, we established in Chapter 3 that the optimal policy is characterized by a sequence of thresholds. Interestingly, there we also identified cases in which the optimal policy is to either inspect the whole batch or not inspect at all. This provides some justification for the inspection mechanism adopted here.

5
Coordinated Production-Inspection in a Tandem System

The central issue in this chapter is the coordination between inspection and production–provided both operations have to be accomplished by the same operator or both consume the same production capacity. To study this issue, we consider a tandem system of two stages, focusing on the control of the first stage, with the second stage representing the aggregate of the rest of the production facility. There is a single server at the first stage that is engaged in processing an item, inspecting the produced item, or staying idle.

Traditional tandem (or, serial) queue models are "make-to-order" systems: an order arrives at the system, goes through the stages in sequence to get processed, and leaves the last stage as an end product (i.e., completed order). Many production systems, however, operate in a "make-to-stock" mode. One simple and popular scheme is the "base-stock control": at every stage there is a target inventory level—the base-stock level, and production (i.e., service) is activated as soon as the finished-goods inventory (at that stage) drops below the base-stock level, regardless of whether there is an order arrival. On the other hand, production at a stage is temporarily suspended when the finished-goods inventory has reached the base-stock level.

Indeed a case can be made that this kind of "inventory queue"—a queueing system operating under some inventory control mechanism—constitutes the basic building block in modeling a production-inventory system, because it captures both the resource contention and the production-inventory dynamics in such systems. Our objective here is to add to this basic model the feature of quality control. In addition to finding the optimal coordination between production and inspection, we will examine whether this optimal policy results in a buffer capacity limit, such that the popular production-inventory control schemes mentioned earlier still apply.

We formulate a semi-Markov decision program (SMDP), with a long-run average cost objective and derive the stationary optimal policy to control and coordinate the production, inspection, and idling processes at the first stage. We show that there exists a threshold value i^*, such that under the optimal policy, once the threshold is reached, production should be suspended at the first stage. This leads naturally to $i^* + 1$ being the required buffer capacity between the two stages.

In §5.1, we start with a detailed problem statement, followed by the SMDP formulation. In §5.2, we focus on establishing the existence and characteristics of stationary optimal policies. In §5.3 we study the structural properties of the optimal policy and establish in particular the buffer capacity limit.

5.1   A Two-Stage Tandem Queue

5.1.1   Problem Statement

We study a tandem system that consists of two stages in series, denoted M_1 and M_2. The first stage, M_1, takes input—raw materials—from an infinite source (a large warehouse, for example) and processes the raw materials into semiproducts, which are then passed on to the second stage, M_2. We use M_2 as an aggregate for all the downstream facilities. It can include, for example, final inspection and packaging, as well as production, but we do not explicitly model these individual processes. There is a buffer between the two stages, and M_2 draws work from the buffer. There is no a priori limit on the buffer capacity (see further discussions in §5.3, in particular, Proposition 5.14).

The decisions to be made are all associated with M_1. Specifically, after completing each item, we have to decide whether to inspect the item. If the item is inspected and found defective, it will be removed from the system—sent to a separate rework facility, for instance. Otherwise, i.e., either the decision is not to inspect the item or the item is found non-defective after inspection, the item proceeds to M_2, from which it leaves the system. After the inspection decision is carried out, regardless of whether the item is inspected, we still have to decide whether to process another item at M_1 or to keep it idle for a period of time.

We will use 0 and 1 to index, respectively, inspection and processing at M_1. Let the random variables X_0 and X_1 denote the generic times for inspecting and processing each item at M_1, and assume that these random variables follow general distributions. Let Y denote the generic processing times at M_2, which follow an exponential distribution. Suppose the processing times (at both stages) and the inspection times are independent, and independent among all the items. Assume the defective rate at M_1 is θ, a given constant.

We need two technical conditions:

(i) the traffic condition: $\mathsf{E}[X_1] + \mathsf{E}[X_0] > \mathsf{E}[Y](1 - \theta)$. That is, at least when M_1 inspects all items, M_2 will have enough capacity to handle the input traffic.

(ii) a "nonexplosiveness" condition:

$$\mathsf{P}[X_1 \geq \delta] \geq q, \quad \mathsf{P}[X_0 \geq \delta] \geq q, \quad \mathsf{P}[Y \geq \delta] \geq q, \tag{5.1}$$

for some constants $\delta > 0$ and $q > 0$. This ensures that, with probability 1, the number of actions taken in any finite time interval will be finite.

(Note that because Y follows an exponential distribution, $\mathsf{P}[Y \geq \delta] \geq q$ is equivalent to $\delta \leq (-\log q)\mathsf{E}[Y]$, which is always true when δ and q are small enough and $\mathsf{E}[Y] > 0$. Hence, condition (ii) only requires that the first two inequalities in (5.1) hold for some small positive constants δ and q.)

The cost data are as follows: the production costs per item at the two stages are C_{P1} and C_{P2}. Inspection at M_1 costs C_I per item, and each defective item passed on to M_2 costs C_D. A good item from M_1 to M_2, on the other hand, earns a revenue of R. There is an inventory holding cost of h per time unit for each item held in the buffer (between the two stages).

To avoid trivial solutions, we assume the following relations on the cost data:

$$C_{P1} + C_{P2} \leq R, \qquad C_I \leq \theta(C_{P2} + C_D). \tag{5.2}$$

Obviously, the first inequality gives incentive for processing at the two stages, and the second one gives incentive for inspection.

5.1.2 The SMDP Formulation

We now formulate the decision problem at M_1 as a semi-Markov decision process. Briefly, the SMDP works as follows: there is a state space, denoted as S; and for each state $z \in S$, $A(z)$ is a set of control actions. Suppose in state z we choose action $a \in A(z)$; we will then have to pay a cost $c(z, a)$, and the sojourn time at z will be $\tau(z, a)$. After the sojourn time expires, the system transits to another state z' with probability $p_{z,z'}(a)$; and the process is repeated. (Refer to Ross [76] for more details.)

Similar to the discrete-time MDP in §4.2.1, a policy $\pi = \{\pi_0, \pi_1, \cdots\}$ is a sequence of rules that prescribes in each decision epoch what action to take. We can similarly define Markovian policy and stationary policy for the SMDP.

Let $\hat{Z}_t$ and $\hat{a}_t$ denote the tth state visited and the action taken in that state. Then, starting from an initial state z, the long-run average cost under policy π is:

$$\bar{V}_\pi(z) = \lim_{n \to \infty} \frac{\mathsf{E}_\pi[\sum_{t=0}^{n} c(\hat{Z}_t, \hat{a}_t)|\hat{Z}_0 = z]}{\mathsf{E}_\pi[\sum_{t=0}^{n} \tau(\hat{Z}_t, \hat{a}_t)|\hat{Z}_0 = z]} \qquad \text{for } z \in S \qquad (5.3)$$

where E_π is the expectation with respect to the probability measure corresponding to policy π.

A policy π^* is said to be optimal if $\bar{V}_{\pi^*}(z) \leq \inf_\pi \bar{V}_\pi(z)$, for any state $z \in S$.

For our control problem, we say the system is in state $(i, 1)$, if M_1 has just completed processing an item, and there are i items in M_2, *excluding* the one that M_1 has just completed. We say the system is in state $(i, 0)$, if M_1 has just completed inspecting an item or has been in idle for a while and there are i items in M_2, including any item that has just passed the inspection at M_1 and subsequently entered M_2. Hence, the state space S is:

$$S = \{(i, 1) : i \geq 0\} \cup \{(i, 0) : i \geq 0\}.$$

In state $(i, 1)$, there are three possible actions: to inspect the item M_1 has just completed, to process another item, or to keep M_1 idle, denoted as a_0, a_1, and a_2 respectively. Note that in the last two cases, the item M_1 has just completed is immediately passed on to M_2, whereas in the first (inspection) case, the item will only be sent to M_2 if and when it passes the inspection. In state $(i, 0)$, there are two possible actions: a_1 and a_2. Inspection is not an option here, because the 0 component of the state indicates that a decision as to whether to inspect has just been made. (Note that the idling action also implies foregoing inspection.)

When action a_2 is taken, we need to specify *how long* to keep M_1 idle. Hence, we shall use a_2^T to denote the idling action that keeps M_1 idle for T units of time. Here we allow T to be equal to

- either X_0, X_1, or Y,

- or a deterministic bounded quantity, $t \in [\delta, L]$ (where $\delta > 0$ is the small positive constant that fulfills the nonexplosiveness condition in (5.1) and L is a large value),

- or any convex combination of the preceding.

This allows a wide range of possibilities while still maintaining the compactness of the action space.

To derive the one-step transition probabilities, the key is to quantify the number of items M_2 completes while M_1 is engaged in a specific action. Due to the i.i.d. (identical and independently distributed) exponential processing times at M_2, the service completion (counting) process from M_2 (given sufficient supply from M_1) follows a Poisson process with rate $1/\,\mathsf{E}[Y]$, denoted as $\{N(t)\}$. In particular, $N(X)$ denotes a Poisson random variable with mean $X/\mathsf{E}[Y]$, where X itself can be a (generic) random variable. For example, when $X = X_1$ or X_0, the time to process or inspect an item at M_1, $N(X)$ denotes the number of items completed at M_2 over that time period. We can now express the one-step transition probabilities as follows:

transition probabilities from state $(i, 1)$:

$$
\begin{aligned}
p_{(i,1)(j,0)}(a_2^T) &= \mathsf{P}[N(T) = i+1-j], \quad i+1 \geq j \geq 1; \\
p_{(i,1)(0,0)}(a_2^T) &= \mathsf{P}[N(T) \geq i+1]; \\
p_{(i,1)(j,1)}(a_1) &= \mathsf{P}[N(X_1) = i+1-j], \quad i+1 \geq j \geq 1; \\
p_{(i,1)(0,1)}(a_1) &= \mathsf{P}[N(X_1) \geq i+1]; \\
p_{(i,1)(j,0)}(a_0) &= \theta \mathsf{P}[N(X_0) = i-j] \\
&\quad + (1-\theta)\mathsf{P}[N(X_0) = i+1-j], \quad i \geq j \geq 1; \\
p_{(i,1)(i+1,0)}(a_0) &= (1-\theta)\mathsf{P}[N(X_0) = 0]; \\
p_{(i,1)(0,0)}(a_0) &= \theta \mathsf{P}[N(X_0) \geq i];
\end{aligned}
$$

transition probabilities from state $(i, 0)$:

$$
\begin{aligned}
p_{(i,0)(j,0)}(a_2^T) &= \mathsf{P}[N(T) = i-j], \quad i \geq j \geq 1; \\
p_{(i,0)(0,0)}(a_2^T) &= \mathsf{P}[N(T) \geq i]; \\
p_{(i,0)(j,1)}(a_1) &= \mathsf{P}[N(X_1) = i-j], \quad i \geq j \geq 1; \\
p_{(i,0)(0,1)}(a_1) &= \mathsf{P}[N(X_1) \geq i].
\end{aligned}
$$

Note the difference between $i + 1$ in the transition probabilities from $(i, 1)$ and i in the transition probabilities from $(i, 0)$, which follows from the difference in the i component of the state definition: the i in $(i, 1)$ excludes the item that has just been completed by M_1, whereas the i in $(i, 0)$ includes the item, if any, that has passed the inspection at M_1 or was sent down to M_2 without inspection.

The sojourn times are as follows: $\tau(z, a_2^T) = T$; while $\tau(z, a_1)$ and $\tau(z, a_0)$ are equal in distributions to X_1 and X_0, respectively.

To derive the one-step cost functions, note the following:

(i) in state $(i, 1)$, both actions a_1 and a_2 result in an item being passed down to M_2 without inspection;

(ii) any item that enters M_2 will be processed there, and hence will incur the production cost C_{P2}; and

(iii) in state $(i, 0)$, no item is sent to M_2 under either action a_1 or a_2.

Hence, the one-step cost functions are:

$$\begin{aligned}
c[(i,1),a_2^T] &= C_{P_2} + \theta C_D - (1-\theta)R + \mathsf{E}[H(i+1,T)]; \\
c[(i,1),a_1] &= C_{P1} + C_{P2} + \theta C_D - (1-\theta)R + \mathsf{E}[H(i+1,X_1)]; \\
c[(i,1),a_0] &= C_I + (1-\theta)C_{P2} - (1-\theta)R + \mathsf{E}[H(i,X_0)]; \\
c[(i,0),a_2^T] &= \mathsf{E}[H(i,T)]; \text{ and} \\
c[(i,0),a_1] &= C_{P1} + \mathsf{E}[H(i,X_1)]. \tag{5.4}
\end{aligned}$$

Here $H(i,t)$ denotes the holding cost in the buffer over a period of t time units, starting with i items there and assuming that no new item is sent to M_1 during this period. It can be derived as follows: let $h_n(i,t)$ denote $H(i,t)$ conditioning on exactly n items being completed by M_2 during t time units. Note that under this conditioning, for $n \leq i-1$, the unordered departure epochs from M_2 of these n items follow the i.i.d. uniform distribution over $(0,t)$. Hence,

$$h_n(i,t) = iht - n(ht/2) = ht(i - n/2), \qquad \text{for } n \leq i-1.$$

For $n = i$, let $\mathcal{T}_i$ denote the time until the departure of the last (i.e., the ith) item. Clearly, $\mathcal{T}_i$ follows an Erlang$-i$ distribution, and we have

$$\begin{aligned}
h_i(i,t) &= \int_s i(hs/2)\mathsf{P}[\mathcal{T}_i \in ds | N(t) \geq i] \\
&= \frac{\frac{ih}{2}\int_0^t s \frac{\lambda^i s^{i-1}}{(i-1)!}e^{-\lambda s}ds}{\mathsf{P}[N(t) \geq i]},
\end{aligned}$$

where $\lambda := 1/\mathsf{E}[Y]$ is the processing rate in M_2. Therefore,

$$\begin{aligned}
H(i,t) &= \sum_{n=0}^{i-1} h_n(i,t)\mathsf{P}[N(t)=n] + h_i(i,t)\mathsf{P}[N(t) \geq i] \\
&= ht\sum_{n=0}^{i-1}(i - \frac{n}{2})\mathsf{P}[N(t)=n] + \frac{ih}{2}\int_0^t \frac{e^{-\lambda s}(\lambda s)^i}{(i-1)!}ds. \tag{5.5}
\end{aligned}$$

The objective function now follows the expression in (5.3); and we want to find an optimal policy π that minimizes the objective function for any (given) initial state $z \in S$.

To solve this SMDP, we first modify the formulation slightly as follows. Note that for the q and δ in (5.1),

$$(R - C_{P2})\mathsf{E}[\tau(z,a)]/q\delta \geq R - C_{P2}$$

because $\mathsf{E}[\tau(z,a)] \geq q\delta$ for all z and a, and $R - C_{P2} \geq 0$. It is also easy to see from (5.4) that

$$c(z,a) \geq C_{P_2} - R \qquad \text{for all } z \text{ and } a.$$

Hence, if we add to each of the one-step cost functions in (5.4) the following quantity

$$(R - C_{P2})\mathsf{E}[\tau(z, a)]/q\delta,$$

the modified one-step cost functions are all nonnegative. Note that with this modification, the objective function in (5.3) becomes

$$\lim_{n \to \infty} \frac{\mathsf{E}_\pi\{\sum_{t=0}^n [c(\hat{Z}_t, \hat{a}_t) + (R - C_{P2})\mathsf{E}[\tau(\hat{Z}_t, \hat{a}_t)]/q\delta] \mid \hat{Z}_0 = i\}}{\mathsf{E}_\pi[\sum_{t=0}^n \tau(\hat{Z}_t, \hat{a}_t) | \hat{Z}_0 = i]}$$

$$= \bar{V}_\pi(i) + (R - C_{P2})/q\delta.$$

Because the last term is independent of the policy π, its inclusion will not affect the solution in any way. But nonnegative one-step cost functions will greatly simplify our discussion later. Hence, we will replace the one-step cost functions, $c(z, a)$, by

$$c(z, a) + (R - C_{P2})\mathsf{E}[\tau(z, a)]/q\delta.$$

For convenience, we will continue to denote the modified one-step cost function as $c(z, a)$ and continue to denote the modified objective function as $\bar{V}_\pi(i)$.

We conclude this section with a lemma on the one-step cost functions.

Lemma 5.1 The one-step cost function, $c(z, a)$, for $z = (i, k)$, is increasing in i, for $k = 0, 1$, and for $a \in A(z)$.

Proof. From the specification of the one-step cost functions, it suffices to show that $H(i, t)$ is increasing in i, for any given $t \geq 0$.

Let $U_i(s)$ denote the number of items in the buffer at time s for $0 \leq s \leq t$, staring with i items there and assuming that no new item joins the buffer. Then $\{U_i(s), 0 \leq s \leq t\}$ is a pure death process with a constant death rate $\lambda = 1/\mathsf{E}[Y]$, and

$$H(i, t) = \mathsf{E}[\int_0^t hU_i(s)ds].$$

Our conclusion is obtained from the fact that $U_i(s)$ is increasing in i in a sample-path sense. $\square$

Remark 5.2 A stronger result on the pure death process $\{U_i(s)\}$ can be found in §2.4. Specifically, from the results there, we know that it is not only increasing but also convex in i; furthermore, it is decreasing and convex in t and submodular in (i, t). These can be readily translated into related properties of $H(i, t)$.

5.2 Stationary Optimal Policies

Write $\bar{\tau}(z,a) := \mathsf{E}[\tau(z,a)]$ for $a \in A(z)$. To establish the existence of a stationary optimal policy, following the standard SMDP theory, here we will prove:

(a) there exists a constant g and a function $v(z)$ satisfying the following relation:

$$v(z) = \min_{a \in A(z)} \left\{ c(z,a) - g\bar{\tau}(z,a) + \sum_{z'} p_{z,z'}(a)v(z') \right\} \qquad (5.6)$$

(b) a policy that prescribes actions achieving the minimum on the right-hand side of (5.6) is optimal.

The equation in (5.6) is known as the optimality equation for the SMDP with a long-run average-cost criterion; see, e.g., Ross [76].

The preceding constitutes Theorems 5.7 and 5.9, the two main results in this section. Our approach is to transform the SMDP into an equivalent discrete-time MDP using the standard transformation technique and then to establish the results through the MDP. (See, for example, Schweitzer [78] or Serfozo [79] for general results on the transformation.) In studying the MDP, we also examine its counterpart under a discounted objective.

Let $\bar{\tau}_{\min}$ be any positive constant such that

$$\bar{\tau}_{\min} \leq \min\{\mathsf{E}[X_1], \mathsf{E}[X_0], \mathsf{E}[Y]\}.$$

Define

$$\bar{c}[z,a] := c[z,a]/\bar{\tau}[z,a] \qquad (5.7)$$

and

$$\bar{p}_{z,z'}(a) := \delta_{zz'} + [p_{z,z'}(a) - \delta_{zz'}]\bar{\tau}_{\min}/\bar{\tau}[z,a], \qquad (5.8)$$

where $\delta_{zz'} = 1$ or 0 corresponding to $z = z'$ or $z \neq z'$.

Lemma 5.3 For any $z \in S$ and $a \in A(z)$,

$$0 \leq \bar{p}_{z,z'}(a) \leq 1 \qquad \text{and} \qquad \sum_{z'} \bar{p}_{z,z'}(a) = 1.$$

Hence, (5.8) defines a set of transition probabilities.

Proof. We only need to show that $\bar{p}_{z,z}(a) \geq 0$, all other relations being trivial. That is, we want to prove

$$1 + \frac{\bar{\tau}_{\min}}{\bar{\tau}[z,a]}[p_{z,z}(a) - 1] \geq 0.$$

If $p_{z,z}(a) = 1$ (e.g., $z = (0,0)$ and $a = a_2$), then the preceding is trivial. Suppose $p_{z,z}(a) < 1$. Then it suffices to prove

$$\bar{\tau}_{\min} \leq \frac{\bar{\tau}[z,a]}{1 - p_{z,z}(a)} \qquad \text{for } a \in A(z). \tag{5.9}$$

Because $\bar{\tau}[z,a_1] = \mathsf{E}[X_1]$ and $\bar{\tau}[z,a_0] = \mathsf{E}[X_0]$, (5.9) obviously holds for $a = a_1$ and $a = a_0$. For $a = a_2^T$, we have $\bar{\tau}[z,a] = \mathsf{E}(T)$ and

$$\frac{\bar{\tau}[z,a]}{1 - p_{z,z}(a)} = \frac{\mathsf{E}(T)}{\mathsf{P}[N(T) \geq 1]} = \frac{\mathsf{E}(T)}{\mathsf{E}[1 - e^{-T/\mathsf{E}[Y]}]} \geq \mathsf{E}[Y],$$

where the inequality follows from $T \geq \mathsf{E}(Y)[1 - e^{-T/\mathsf{E}[Y]}]$, for any $T \geq 0$. (This, in turn, is due to $x \geq 1 - e^{-x}$ for $x \geq 0$, with $x = T/\mathsf{E}[Y]$.) Hence, (5.9) also holds when $a = a_2^T$. $\square$

Now, consider the discrete-time MDP with one-step cost functions $\bar{c}[z,a]$ and transition probabilities $\bar{p}_{z,z'}(a)$ as specified earlier. Denote the long-run average objective function for this MDP as $\hat{V}_\pi(z)$. The following lemma enables us to discuss the original SMDP through the discrete-time MDP.

Proposition 5.4 The statements in (a) and (b) are equivalent.

(a) There exists a constant $\bar{g}$ and a nonnegative function $\bar{v}(z)$ that solve the following discrete-time MDP optimality equation:

$$\bar{v}(z) = \min_{a \in A(z)} \left\{ \bar{c}(z,a) - \bar{g} + \sum_{z'} \bar{p}_{z,z'}(a)\bar{v}(z') \right\}. \tag{5.10}$$

(b) There exists a constant g and a nonnegative function $v(z)$ that solve the optimality equation in (5.6) for the original SMDP.

Furthermore, we have $g = \bar{g}$ and $v(z) = \bar{\tau}_{\min}\bar{v}(z)$.

Proof. It is straightforward to show that if $\bar{g}$ and $\bar{v}(z)$ solve (5.10), then $g = \bar{g}$ and $v(z) = \bar{\tau}_{\min}\bar{v}(z)$ solve (5.6); and vice versa. $\square$

Therefore, it suffices to establish (a) in Proposition 5.4. To this end, consider the discounted-cost version of the MDP, with the following objective function

$$V_{\alpha,\pi}(z) = \mathsf{E}_\pi \left[\sum_{t=0}^{\infty} e^{-\alpha t} \bar{c}(\hat{Z}_t, \hat{a}_t) | \hat{Z}_0 = z \right],$$

where $\alpha \in (0,1)$ is the discount factor. Let $V_\alpha(z) = \inf_\pi V_{\alpha,\pi}(z)$ denote the optimal discounted objective function. From Lemma 4.2 we have

$$V_\alpha(z) = \min_a \left\{ \bar{c}(z,a) + e^{-\alpha} \sum_{z'} \bar{p}_{z,z'}(a)V_\alpha(z') \right\}, \qquad z \in S. \tag{5.11}$$

Lemma 5.5 For $z = (i, k)$, $V_\alpha(z)$ is increasing in i, for $k = 0, 1$ and for any given $\alpha \in (0, 1)$.

Proof. For $z = (i, k)$ and given $\alpha > 0$, define

$$
\begin{aligned}
v_0(z) &= 0, \\
v_n(z) &= \min_a \Big\{ \bar{c}(z, a) + e^{-\alpha} \sum_{z'} \bar{p}_{z,z'}(a) v_{n-1}(z') \Big\}.
\end{aligned}
\tag{5.12}
$$

From Lemma 4.2, we have

$$
v_n(z) \to V_\alpha(z) \qquad \text{as } n \to \infty.
$$

Hence, it suffices to show that $v_n(z)$ is increasing in i for any n. Use induction. For $n = 0$, this is trivial because $v_0(z) = 0$. Suppose $v_{n-1}(z)$ is increasing in i for some $n \geq 1$. Note that with $\hat{Z}_n = (i_n, k_n)$, $[i_n | \hat{Z}_{n-1} = z, a_{n-1} = a]$ is stochastically increasing in i, while k_n is independent of i. This implies that

$$
\sum_{z'} \bar{p}_{z,z'}(a) v_{n-1}(z') = \mathsf{E}[v_{n-1}(\hat{Z}_n) | \hat{Z}_{n-1} = z, a_{n-1} = a]
$$

is increasing in i, as $v_{n-1}(\cdot)$ is increasing (induction hypothesis). In addition, from Lemma 5.1, we know $c(z, a)$ is increasing in i; and because $\bar{\tau}(z, a)$ is independent of z, $\bar{c}(i, a)$ is increasing in i. Hence, from (5.12), $v_n(z)$ is also increasing in i. $\square$

Lemma 5.6 For the MDP, we have the following results:

(i) there exists a stationary policy π under which the average cost is finite; and

(ii) it is possible (i.e., under certain policy) to go from any given state $z \in S$ to state $(0, 0)$ or to state $(0, 1)$ with a finite expected total cost.

Proof. (i) Let π be a policy under which we always choose action a_2^T at each decision epoch, for a positive constant $T \geq q\delta$ (q and δ are the positive constants in the nonexplosiveness condition in (5.1)). Suppose $z = (i, k)$ is the initial state. Because the processing time for each item at M_2 is exponentially distributed, the time until M_2 completing all i items follows an Erlang-i distribution, which has a finite expectation. This implies that, under policy π, starting from state z, the system will reach state $(0, 0)$ in a finite number of transitions (in expectation) and then stay there forever. Therefore, the expected average cost under π is

$$
\hat{V}_\pi(z) = \bar{c}[(0, 0), a_2^T] = c[(0, 0), a_2^T] / \mathsf{E}(T) \leq (R - C_{P2}) / q\delta,
$$

which is finite.

(ii) With the policy π specified earlier, we can reach $(0,0)$ in a finite number of steps (in expectation). Note that from (5.4) and (5.7), if $z = (i,0)$, then the cost in each step is

$$\bar{c}[(j,0), a_2^T] = (R - C_{P2})/q\delta + \mathsf{E}[H(j,T)]/T,$$

for some $j \leq i$, and is hence bounded by $(R - C_{P2})/q\delta + ih$. On the other hand, if $z = (i,1)$, then the cost in the first transition is finite, and after that the one-step costs become what we just specified earlier, until the state $(0,0)$ is reached. Hence, the expected total cost going from z to $(0,0)$ is finite. The cost going from $(0,0)$ to $(0,1)$ is also finite; hence we have the desired conclusion. $\square$

We are now ready to present the main result of this section.

Theorem 5.7 Consider the SMDP. There exists a constant g and a function $v(z)$ for $z \in S$ satisfying the optimality equation (5.6). Furthermore, $v(z) = v(i,k)$ is increasing in i (for $k = 0$ or 1).

Proof. From Proposition 5.4, it suffices to show the desired conclusions for the discrete-time MDP. Specifically, we want to show that there exist g and $\bar{v}(z)$ satisfying (5.10), and that $\bar{v}(z)$ is increasing in i for $z = (i,k)$.

First, from Lemma 5.6, there exists a stationary policy π under which we have a finite average cost, denoted as $\gamma = \hat{V}_\pi(0,0) < \infty$. Without loss of generality, suppose for this π, there exists a sequence of discount factors $\alpha \to 0$, such that
$$V_{\alpha,\pi}(0,0) \leq V_{\alpha,\pi}(0,1); \tag{5.13}$$

we will only focus on these αs. Otherwise, i.e., if the inequality in the opposite direction, holds for an infinite sequence of αs, we can use $(0,1)$ in lieu of $(0,0)$ in the following discussion.

Let $\underline{0}$ denote the state $(0,0)$ and let $M(z, \underline{0})$ denote the minimum expected cost going from state z to state $\underline{0}$. From Lemma 5.6, $M(z, \underline{0}) < \infty$. Because the one-step cost is non-negative, for any discount factor α, we have

$$V_\alpha(z) \leq M(z, \underline{0}) + V_\alpha(\underline{0}).$$

On the other hand, for $z = (i,k)$, making use of the increasing property in Lemma 5.5, we have

$$
\begin{aligned}
& V_\alpha(i,k) - V_\alpha(0,0) \\
= \ & V_\alpha(i,k) - V_\alpha(0,k) + V_\alpha(0,k) - V_\alpha(0,0) \\
\geq \ & V_\alpha(0,k) - V_\alpha(0,0),
\end{aligned}
$$

which is bounded from below for any α, due to the fact that the cost for going from $(0,0)$ to $(0,k)$ (for $k = 0,1$) is finite. Therefore, for $z = (i,k)$,

$k = 0, 1$, $|V_\alpha(z) - V_\alpha(\underline{0})|$ is bounded (the bound may depend on z but is independent of α).

This implies that there exists a subsequence of discount factors $\{\alpha_n\}$ with $\alpha_n \to 0$, such that the difference, $V_{\alpha_n}(z) - V_{\alpha_n}(\underline{0})$, is convergent. In what follows, we will show that $\alpha V_\alpha(\underline{0})$ is also bounded, and consequently, we can find a subsequence $\{\alpha_n\}$ such that $\alpha_n V_{\alpha_n}(\underline{0})$ is also convergent.

By the definition of $V_\alpha(\underline{0})$, $V_\alpha(\underline{0}) \le V_{\alpha,\pi}(\underline{0})$. We will next show that $V_{\alpha,\pi}(\underline{0}) \le \gamma/\alpha$, and hence $\alpha V_\alpha(\underline{0}) \le \gamma$. To this end, reason as follows. Starting from state $\underline{0}$, suppose we have opportunities to restart our system from state $\underline{0}$, and the opportunities occur randomly following a Poisson process with rate α. Let $\hat{V}'_\pi(\underline{0})$ denote the long-run average cost if we follow the stationary policy π and restart the system every time the opportunity occurs. Let τ denote the time between two such opportunities, which is exponentially distributed with a parameter (rate) α. The (nondiscounted) cost between two such opportunities is

$$\mathsf{E}_\pi[\sum_{t=0}^{\lfloor \tau \rfloor} \bar{c}(\hat{Z}_t, \hat{a}_t)|\hat{Z}_0 = \underline{0}] = \mathsf{E}_\pi[\sum_{t=0}^{\infty} \bar{c}(\hat{Z}_t, \hat{a}_t)\mathsf{P}[\tau > t]|\hat{Z}_0 = \underline{0}]$$

$$= \mathsf{E}_\pi[\sum_{t=0}^{\infty} e^{-t\alpha}\bar{c}(\hat{Z}_t, \hat{a}_t)|\hat{Z}_0 = \underline{0}]$$

$$= V_{\alpha,\pi}(\underline{0}).$$

Here $\lfloor \tau \rfloor$ denotes the integer part (i.e., lower floor) of τ. From the theory of renewal processes (e.g., Ross [73]), we have

$$\hat{V}'_\pi(\underline{0}) = \mathsf{E}_\pi[\sum_{t=0}^{\lfloor \tau \rfloor} \bar{c}(\hat{Z}_t, \hat{a}_t)|\hat{Z}_0 = \underline{0}]/\mathsf{E}[\tau] = \alpha V_{\alpha,\pi}(\underline{0}). \tag{5.14}$$

On the other hand, note that $\hat{V}'_\pi(\underline{0})$ is the average cost operating under π making use of all the restarting opportunities, while $\gamma = \hat{V}_\pi(\underline{0})$ is the average cost operating under π without any restarting opportunity. We have $\gamma \ge \hat{V}'_\pi(\underline{0})$, due to the following reasoning: Because the one-step cost function is increasing in i (Lemma 5.1), if we can reset the state from (i, k) to $(0, k)$, then the cost is reduced. Furthermore, from the assumption in (5.13), to restart at $(0, 0)$ is better than to restart at $(0, 1)$, and hence, making use of all the opportunities to restart from $(0, 0)$ is better than making no use of such opportunities. Therefore, $V_{\alpha,\pi}(\underline{0}) \le \gamma/\alpha$ follows from (5.14).

Because both $V_\alpha(z) - V_\alpha(\underline{0})$ and $\alpha V_\alpha(\underline{0})$ are bounded for an infinite sequence of $\alpha \to 0$ through a diagonalization argument, we can find a sequence of discount factors $\{\alpha_n\}$ such that as $\alpha_n \to 0$,

$$V_{\alpha_n}(z) - V_{\alpha_n}(\underline{0}) \to \bar{v}(z) \quad \text{and} \quad \alpha_n V_{\alpha_n}(\underline{0}) \to g,$$

for some constant g and some function $\bar{v}(z)$.

From (5.11) we have

$$V_\alpha(z) - V_\alpha(\underline{0})$$

$$= \min_a \left\{ \bar{c}(z, a) + e^{-\alpha} \sum_{z'} \bar{p}_{z,z'}(a)[V_\alpha(z') - V_\alpha(\underline{0})] - (1 - e^{-\alpha})V_\alpha(\underline{0}) \right\}.$$

Replacing α by α_n in the preceding equation and letting $\alpha_n \to 0$, we obtain

$$\bar{v}(z) = \min_a \left\{ \bar{c}(z, a) - g + \sum_{z'} \bar{p}_{z,z'}(a)\bar{v}(z') \right\}.$$

(Here we have used the fact that $(1 - e^{-\alpha})V_\alpha(\underline{0}) = \alpha V_\alpha(\underline{0}) + o(\alpha)$ for small α.) Hence, g and $\bar{v}(z)$ satisfy the optimality equation in (5.10). Because, from Lemma 5.5, $V_\alpha(z)$ is increasing in i, $\bar{v}(z)$ is also increasing in i, and so is $v(z) = \bar{\tau}_{\min}\bar{v}(z)$. Furthermore, from Proposition 5.4, g and $v(z)$ satisfy the optimality equation in (5.6). $\square$

Lemma 5.8 For any policy π, if $\bar{V}_\pi(z)$ is finite, then

$$\lim_{t \to \infty} \mathsf{E}_\pi[v(\hat{Z}_t)|\hat{Z}_0 = z] < \infty.$$

Proof. Use contradiction. Suppose $\mathsf{E}_\pi[v(\hat{Z}_t)|\hat{Z}_0 = z] \to \infty$, as $t \to \infty$. We will show that $\bar{V}_\pi(z) \geq R$ for any given positive constant R, contradicting the finiteness assumed of $\bar{V}_\pi(z)$.

From Theorem 5.7,

$$v(z) = \bar{\tau}_{\min} \lim_{\alpha \to 0}[V_\alpha(z) - V_\alpha(\underline{0})] \leq \bar{\tau}_{\min}M(z, \underline{0}),$$

where $M(z, \underline{0})$ is the cost associated with going from z to $\underline{0}$. Hence,

$$\mathsf{E}_\pi[M(\hat{Z}_t, \underline{0})|\hat{Z}_0 = z] \to \infty \qquad \text{as} \qquad t \to \infty.$$

This implies, with $\hat{Z}_t = (i_t, k_t)$, that $i_t \to \infty$ with a positive probability $p > 0$. That is, for any positive integer R, there exists an integer $n(R)$ such that $i_t \geq R$ with probability $p > 0$ for any $t \geq n(R)$; and hence, from that point onward, the expected holding cost per time unit is at least pR. Therefore, for $t \geq n(R)$,

$$\mathsf{E}_\pi[c(\hat{Z}_t, \hat{a}_t)] \geq pR\mathsf{E}[\tau(\hat{Z}_t, \hat{a}_t)];$$

and

$$\bar{V}_\pi(z) = \lim_{m \to \infty} \frac{\mathsf{E}[\sum_{t=0}^m c(\hat{Z}_t, \hat{a}_t)]}{\mathsf{E}[\sum_{t=0}^m \tau(\hat{Z}_t, \hat{a}_t)]}$$

$$= \lim_{m \to \infty} \frac{\mathsf{E}[\sum_{t=0}^{n(R)-1} c(\hat{Z}_t, \hat{a}_t) + \sum_{t=n(R)}^m c(\hat{Z}_t, \hat{a}_t)]}{\mathsf{E}[\sum_{t=0}^m \tau(\hat{Z}_t, \hat{a}_t)]}$$

$$\geq \lim_{m \to \infty} \frac{\mathsf{E}[\sum_{t=0}^{n(R)-1} c(\hat{Z}_t, \hat{a}_t)] + \sum_{t=n(R)}^m pR\mathsf{E}[\tau(\hat{Z}_t, \hat{a}_t)]}{\mathsf{E}[\sum_{t=0}^m \tau(\hat{Z}_t, \hat{a}_t)]}$$

$$= pR.$$

Because $p > 0$ and R can be chosen as any positive integer, we have $\bar{V}_\pi(z) = \infty$, a contradiction. $\square$

Theorem 5.9 For the g and $v(z)$ in Theorem 5.7, let π^* be a stationary policy that, for each state z, prescribes an action that achieves the minimum on the right-hand side of (5.6). Then π^* is optimal, and $\bar{V}_{\pi^*}(z) = g$ for any $z \in S$.

Proof. Note that with this π^*,

$$v(z) = c(z, \pi^*(z)) - g\tau(z, \pi^*(z)) + \mathsf{E}_{\pi^*}[v(\hat{Z}_1)|\hat{Z}_0 = z].$$

Using this equation repeatedly we have

$$v(z) = \mathsf{E}_{\pi^*}\Big[\sum_{t=0}^n c(\hat{Z}_t, \hat{a}_t) - g\sum_{t=0}^n \tau(\hat{Z}_t, \hat{a}_t) + v(\hat{Z}_{n+1})|\hat{Z}_0 = z\Big],$$

which implies

$$
\begin{aligned}
g &= \frac{\mathsf{E}_{\pi^*}[\sum_{t=0}^n c(\hat{Z}_t, \hat{a}_t)|\hat{Z}_0 = z]}{\mathsf{E}_{\pi^*}[\sum_{t=0}^n \tau(\hat{Z}_t, \hat{a}_t)|\hat{Z}_0 = z]} - \frac{v(z)}{\mathsf{E}_{\pi^*}[\sum_{t=0}^n \tau(\hat{Z}_t, \hat{a}_t)|\hat{Z}_0 = z]} \\
&\quad + \frac{\mathsf{E}_{\pi^*}[v(\hat{Z}_{n+1})|\hat{Z}_0 = z]}{\mathsf{E}_{\pi^*}[\sum_{t=0}^n \tau(\hat{Z}_t, \hat{a}_t)|\hat{Z}_0 = z]} \\
&\geq \frac{\mathsf{E}_{\pi^*}[\sum_{t=0}^n c(\hat{Z}_t, \hat{a}_t)|\hat{Z}_0 = z]}{\mathsf{E}_{\pi^*}[\sum_{t=0}^n \tau(\hat{Z}_t, \hat{a}_t)|\hat{Z}_0 = z]} - \frac{v(z)}{\mathsf{E}_{\pi^*}[\sum_{t=0}^n \tau(\hat{Z}_t, \hat{a}_t)|\hat{Z}_0 = z]},
\end{aligned}
$$

where the inequality is from the fact that $v(\hat{Z}_{n+1}) \geq 0$. Let $n \to \infty$ we get $g \geq \bar{V}_{\pi^*}(z)$.

Suppose π is any policy that yields a finite average cost. Then similarly, from (5.6), we have

$$v(z) \leq \mathsf{E}_\pi\Big[\sum_{t=0}^n c(\hat{Z}_t, \hat{a}_t) - g\sum_{t=0}^n \tau(\hat{Z}_t, \hat{a}_t) + v(\hat{Z}_{n+1})|\hat{Z}_0 = z\Big],$$

which is equivalent to

$$
\begin{aligned}
g &\leq \frac{\mathsf{E}_\pi[\sum_{t=0}^n c(\hat{Z}_t, \hat{a}_t)|\hat{Z}_0 = z]}{\mathsf{E}_\pi[\sum_{t=0}^n \tau(\hat{Z}_t, \hat{a}_t)|\hat{Z}_0 = z]} - \frac{v(z)}{\mathsf{E}_\pi[\sum_{t=0}^n \tau(\hat{Z}_t, \hat{a}_t)|\hat{Z}_0 = z]} \\
&\quad + \frac{\mathsf{E}_\pi[v(\hat{Z}_{n+1})|\hat{Z}_0 = z]}{\mathsf{E}_\pi[\sum_{t=0}^n \tau(\hat{Z}_t, \hat{a}_t)|\hat{Z}_0 = z]}.
\end{aligned}
\tag{5.15}
$$

From Lemma 5.8, we have

$$\lim_{n\to\infty} \frac{\mathsf{E}_\pi[v(\hat{Z}_{n+1})|\hat{Z}_0 = z]}{\sum_{t=0}^n \mathsf{E}_\pi[\tau(\hat{Z}_t, \hat{a}_t)|\hat{Z}_0 = z]} = 0,$$

because the denominator goes to infinity (due to the nonexplosiveness condition in (5.1)). Note that the first term on the right-hand side of (5.15) converges to $\bar{V}_\pi(z)$. Hence, letting $n \to \infty$ in (5.15) yields $g \leq \bar{V}_\pi(z)$. This, together with $g \geq \bar{V}_{\pi^*}(z)$, implies

$$\bar{V}_{\pi^*}(z) = g = \inf_\pi \bar{V}_\pi(z).$$

Therefore, π^* is optimal. $\square$

5.3   Structure of the Optimal Policy

Following Theorem 5.9, we can derive the optimal action in each state by choosing the action that minimizes the right-hand side of the optimality equation in (5.6). In doing so, however, we must first decide how to select the length of the idle period T in action a_2^T. The following theorem gives an answer to this issue. It turns out that we only need to consider $T = Y$, idle until there is a departure from M_2. (Recall Y follows an exponential distribution and is hence memoryless.)

Theorem 5.10 Under the stationary optimal policy π^*, if it is optimal to take action a_2 (i.e., to keep M_1 idling) in some state z, then it is optimal to continue this action until the next state change (i.e., when an item departs from M_2).

Proof. Suppose $a = a_2^T$ is optimal in state z for some T. Recall Y denotes the service time at M_2. We prove here that a_2^Y is at least as good as a_2^T.

First, suppose $z = (i, 0)$. Then, following Theorem 5.9, we know that a_2^T being optimal in z implies the following:

$$
\begin{aligned}
v(i,0) &= c[(i,0), a_2^T] - g\mathsf{E}[T] + \sum_j p_{(i,0),(j,0)}(a_2^T)v(j,0) \\
&= \min_a \{ c[(i,0), a] - g\bar{\tau}[(i,0), a] + \sum_j p_{(i,0),(j,0)}(a)v(j,0) \},
\end{aligned}
$$

and the relationship remains true as along as there are no state changes. Let T_j and Y_j, $j = 0, 1, 2, \cdots$, be i.i.d. samples of T and Y. Define

$$n_Y = \min\{n : \sum_{j=0}^n T_j > Y\}.$$

Then, it is optimal to apply action a_2^T in state z a total of $n_Y + 1$ times (with i.i.d. samples of T as the idle periods) before the system transits to another state. Let

$$\tau_Y = \sum_{j=0}^{n_Y} T_j \quad \text{and} \quad K = \max\{m \leq i - 1 : \sum_{j=0}^m Y_j \leq \tau_Y\}.$$

Then τ_Y is the total elapsed time after applying $n_Y + 1$ times the action a_2^T; and $K + 1$ is the number of departures from M_2 in τ_Y units of time.

Applying Theorem 5.9 repeatedly, we have

$$
\begin{aligned}
v(i,0) &= c[(i,0),a_2^T] - g\mathsf{E}[T] + \sum_j p_{(i,0),(j,0)}(a_2^T)v(j,0) \\
&= \mathsf{E}[\sum_{j=0}^{n_Y} c(Z_j, a_2^{T_j}) - g\sum_{j=0}^{n_Y} \tau(Z_j, a_2^{T_j}) + v(z_{n_Y+1})|Z_0 = (i,0)] \\
&= \mathsf{E}[\sum_{j=0}^{K}(i-j)hY_j] + \mathsf{E}[(\tau_Y - \sum_{j=0}^{K} Y_j)(i-K)] \\
&\quad + [(R - C_{P2})/(q\delta) - g]\mathsf{E}[\tau_Y] + \mathsf{E}[v(i-K,0)] \\
&\geq \mathsf{E}[\sum_{j=0}^{K}(i-j)hY_j] + [(R - C_{P2})/(q\delta) - g]\mathsf{E}[\sum_{j=0}^{K} Y_j] \\
&\quad + \mathsf{E}[v(i-K,0)], \tag{5.16}
\end{aligned}
$$

where the inequality follows from $g \leq (R - C_{P2})/(q\delta)$, because g is the optimal average cost, while $(R - C_{P2})/(q\delta)$ is the average cost of a policy that always takes only action a_2^T all the time.

On the other hand, applying $K + 1$ times the action a_2^Y, we have

$$
\begin{aligned}
&c[(i,0),a_2^Y] - g\bar{\tau}[(i,0),a_2^Y] + \sum_j p_{(i,0),(j,0)}(a_2^Y)v(j,0) \\
&\leq \mathsf{E}[\sum_{j=0}^{K}(i-j)hY_j] + [(R - C_{P2})/(q\delta) - g]\mathsf{E}[\sum_{j=0}^{K} Y_j] + \mathsf{E}[v(i-K,0)] \\
&\leq c[(i,0),a_2^T] - g\mathsf{E}[T] + \sum_j p_{(i,0),(j,0)}(a_2^T)v(j,0).
\end{aligned}
$$

Here the last equality follows from (5.16). Therefore, it is optimal to take action a_2^Y in state $z = (i,0)$.

Now suppose $z = (i,1)$. Clearly, if it is optimal to take action a_2^T in state $(i,1)$, then it is also optimal to take the same action in state $(i+1,0)$. Hence, similar to the preceding argument, if it is optimal to take action a_2^T in $(i,1)$, then it is optimal to apply the same action n_Y times, until a departure takes place in M_2. A similar discussion then yields the conclusion that a_2^Y is at least as good as a_2^T for any T. □

Next, we want to establish a threshold property of the optimal policy (Theorem 5.13): when the number of jobs in the buffer reaches a certain level, the processing at M_1 should be stopped. We need two lemmas.

Lemma 5.11 For any $j \geq 0$, $v(j+1,0) - v(j,0) \geq jh\mathsf{E}[Y]$.

Proof. Let S_j denote the time needed to complete processing j items in M_2; S_j follows an Erlang-j distribution. For any given policy π, let n_π denote the number of steps needed, starting from state $(j+1, 0)$, until the jth departure from M_2, i.e.,

$$n_\pi = \min\{n : \sum_{t=0}^{n} \tau(\hat{Z}_t, \hat{a}_t) \geq S_j\}$$

given that $\hat{Z}_0 = (j+1, 0)$.

Recall that π^* denote the stationary optimal policy. Suppose $\pi' = (\pi'_0, \pi'_1, \cdots)$ is a (nonstationary) policy defined through π^* by

$$\pi'_n(i, k) = \pi^*(i+1, k), \quad 0 \leq n \leq n_{\pi'}$$

and

$$\pi'_n(i, k) = \pi^*(i, k), \quad n \geq n_{\pi'} + 1$$

for $k = 0, 1$. That is, π' takes the action $\pi^*(i+1, k)$ in state (i, k) until the jth departure from M_2 and then follows exactly π^* after that. Note with π' defined this way, $n_{\pi'}$ and n_{π^*} have the same distribution.

From (5.6), we have, for any policy π and any state z,

$$\begin{aligned} v(z) &\leq c(z, \pi(z)) - g\tau(z, \pi(z)) + \mathsf{E}_\pi v(\hat{Z}_1)|\hat{Z}_0 = z] \\ &\leq \mathsf{E}_\pi[\sum_{t=0}^{n} c(\hat{Z}_t, \hat{a}_t) - g\sum_{t=0}^{n} \tau(\hat{Z}_t, \hat{a}_t) + v(\hat{Z}_{n+1})|\hat{Z}_0 = z] \end{aligned}$$

for $n \geq 1$. And the inequalities will become an equality if $\pi = \pi^*$. Therefore, letting $z = (j+1, 0)$ for the equality (with $\pi = \pi^*$) and $z = (j, 0)$ for the inequality (with $\pi = \pi'$), we have

$$v(j+1, 0) - v(j, 0) \qquad\qquad (5.17)$$

$$\geq \mathsf{E}_{\pi^*}[\sum_{t=0}^{n_{\pi^*}} c(\hat{Z}_t, \hat{a}_t) - g\sum_{t=0}^{n_{\pi^*}} \tau(\hat{Z}_t, \hat{a}_t) + v(\hat{Z}_{n_{\pi^*}+1})|\hat{Z}_0 = (j+1, 0)]$$

$$-\mathsf{E}_{\pi'}[\sum_{t=0}^{n_{\pi'}} c(\hat{Z}_t, \hat{a}_t) - g\sum_{t=0}^{n_{\pi'}} \tau(\hat{Z}_t, \hat{a}_t) + v(\hat{Z}_{n_{\pi'}+1})|\hat{Z}_0 = (j, 0)].$$

Notice that $\tau(z, a)$ depends on a only. Also note that with π' and π^* starting from $(j, 0)$ and $(j+1, 0)$, respectively, the construction of the two policies ensures that exactly the same action will be taken at each decision epoch before the jth departure. Hence,

$$\mathsf{E}_{\pi^*}[\sum_{t=0}^{n_{\pi^*}} \tau(\hat{Z}_t, \hat{a}_t)|\hat{Z}_0 = (j+1, 0)] - \mathsf{E}_{\pi'}[\sum_{t=0}^{n_{\pi'}} \tau(\hat{Z}_t, \hat{a}_t)|\hat{Z}_0 = (j, 0)] = 0.$$

In addition, the process that starts from $(j+1, 0)$ incurs an additional holding cost of h per time unit throughout the period until the jth departure. Hence, we have

$$\mathsf{E}_{\pi^*}[\sum_{t=0}^{n_{\pi^*}} c(\hat{Z}_t, \hat{a}_t)|\hat{Z}_0 = (j+1,0)] - \mathsf{E}_{\pi'}[\sum_{t=0}^{n_{\pi'}} c(\hat{Z}_t, \hat{a}_t)|\hat{Z}_0 = (j,0)]$$
$$\geq h\mathsf{E}[S_j]$$
$$= hj\mathsf{E}[Y]. \tag{5.18}$$

Therefore, from (5.17), to get the desired inequality, it only remains to prove

$$\mathsf{E}_{\pi^*}[v(\hat{Z}_{n_{\pi^*}+1})|\hat{Z}_0 = (j+1,0)] - \mathsf{E}_{\pi'}[v(\hat{Z}_{n_{\pi'}+1})|\hat{Z}_0 = (j,0)] \geq 0. \tag{5.19}$$

Let $(i_{n_{\pi^*}+1}, k)$ denote the state $\hat{Z}_{n_{\pi^*}+1}$ reached under π^* starting from $\hat{Z}_0 = (j+1,0)$; and let $(i_{n_{\pi'}+1}, k)$ denote the state $\hat{Z}_{n_{\pi'}+1}$ reached under π' starting from $\hat{Z}_0 = (j,0)$. Then clearly, $i_{n_{\pi^*}+1} \geq^{st} i_{n_{\pi'}+1}$, which implies (5.19), because $v(\cdot)$ is increasing in i (Theorem 5.7). $\square$

Lemma 5.12 For any $j \geq 0$, $v(j,1) - v(j,0) \geq \Delta(j)$, where

$$\Delta(j) := C_I + (1-\theta)(C_{P2} - R) + (1-\theta)(jh\mathsf{E}[Y] - h\mathsf{E}[X_0]).$$

Proof. We prove the desired inequality in three cases, corresponding to the three possible actions in state $(j,1)$ under policy π^*.

First, suppose $\pi^*(j,1) = a_2^T$ for some T. Define a (nonstationary) policy $\pi' = \{\pi'_0, \pi'_1, \cdots\}$ as follows:

$$\pi'_0(j,0) = a_2^T; \qquad \pi'_n(i,k) = \pi^*(i+1,k), \quad 1 \leq n \leq n_{\pi'};$$

and $\pi'_n(i,k) = \pi^*(i,k)$, for $n \geq n_{\pi'} + 1$. Here n_π is the number of steps until the jth departure from M_2, as defined in Lemma 5.11. Then, similar to (5.17), we have

$$v(j,1) - v(j,0) \tag{5.20}$$
$$\geq \mathsf{E}_{\pi^*}[\sum_{t=0}^{n_{\pi^*}} c(\hat{Z}_t, \hat{a}_t)] - g\sum_{t=0}^{n_{\pi^*}} \tau(\hat{Z}_t, \hat{a}_t)] + v(\hat{Z}_{n_{\pi^*}+1})|\hat{Z}_0 = (j,1)]$$
$$-\mathsf{E}_{\pi'}[\sum_{t=0}^{n_{\pi'}} c(\hat{Z}_t, \hat{a}_t)] - g\sum_{t=0}^{n_{\pi'}} \tau(\hat{Z}_t, \hat{a}_t)] + v(\hat{Z}_{n_{\pi'}+1})|\hat{Z}_0 = (j,0)].$$

Note that by using action a_2^T in state $(j,1)$, the number of items in M_2 changes from j to $j+1$ immediately. As in Lemma 5.11, we have

$$\mathsf{E}_{\pi^*}[\sum_{t=0}^{n_{\pi^*}} \tau(\hat{Z}_t, \hat{a}_t)|\hat{Z}_0 = (j,1)] - \mathsf{E}_{\pi'}[\sum_{t=0}^{n_{\pi'}} \tau(\hat{Z}_t, \hat{a}_t)|\hat{Z}_0 = (j,0)] = 0$$

and

$$\mathsf{E}_{\pi^*}[v(\hat{Z}_{n_{\pi^*}+1})|\hat{Z}_0 = (j,1)] - \mathsf{E}_{\pi'}[v(\hat{Z}_{n_{\pi'}+1})|\hat{Z}_0 = (j,0)] \geq 0.$$

Furthermore,

$$\mathsf{E}_{\pi^*}[\sum_{t=0}^{n_{\pi^*}} c(\hat{Z}_t, \hat{a}_t)|\hat{Z}_0 = (j,1)] - \mathsf{E}_{\pi'}[\sum_{t=0}^{n_{\pi'}} c(\hat{Z}_t, \hat{a}_t)|\hat{Z}_0 = (j,0)]$$

$$= c[(j,1), a_2^T] - c[(j,0), a_2^T] + \mathsf{E}_{\pi^*}[\sum_{t=1}^{n_{\pi^*}} c(\hat{Z}_t, \hat{a}_t)|\hat{Z}_0 = (j,1)]$$

$$-\mathsf{E}_{\pi'}[\sum_{t=1}^{n_{\pi'}} c(\hat{Z}_t, \hat{a}_t)|\hat{Z}_0 = (j,0)]$$

$$= C_{P2} - (1-\theta)R + \theta C_D + H(j+1,T) - H(j,T)$$

$$+\mathsf{E}_{\pi^*}[\sum_{t=1}^{n_{\pi^*}} c(\hat{Z}_t, \hat{a}_t)|\hat{Z}_0 = (j,1)] - \mathsf{E}_{\pi'}[\sum_{t=1}^{n_{\pi'}} c(\hat{Z}_t, \hat{a}_t)|\hat{Z}_0 = (j,0)]$$

$$\geq C_{P2} - (1-\theta)R + \theta C_D + jh\mathsf{E}[Y]$$

$$= \Delta(j) + \theta(C_{P_2} + C_D) - C_I + (1-\theta)h\mathsf{E}[X_0] + \theta jh\mathsf{E}[Y]$$

$$> \Delta(j).$$

Here the first inequality follows from the same argument as in (5.18) and the last inequality follows from the definition of $\Delta(j)$ and the assumption in (5.2) that $C_I \leq \theta(C_{P_2}+C_D)$. Hence, $v(j,1)-v(j,0) > \Delta(j)$, from (5.20).

Second, suppose $\pi^*(j,1) = a_1$. Define the policy π' as follows:

$$\pi_0(j,0) = a_1; \qquad \pi_n(i,k) = \pi^*(i+1,k), \quad 1 \leq n \leq n_{\pi'};$$

and $\pi_n(i,k) = \pi^*(i,k)$, for $n \geq n_{\pi'}$. Then, following the same argument as in the first case, we have

$$v(j,1) - v(j,0) \geq C_{P2} - (1-\theta)R + \theta C_D + jh\mathsf{E}[Y] > \Delta(j).$$

Last, suppose $\pi^*(j,1) = a_0$. Define the policy π' as follows:

$$\pi_0(j,0) = a_2^{X_0}; \qquad \pi_n(i,k) = \pi^*(i,k), \quad n \geq 1.$$

Note here that $a_2^{X_0}$ denotes the action a_2 with an idling time $T = X_0$, a random sample from the distribution of X_0, the inspection time. Then, similar to (5.20), by conditioning on the outcome of the inspection, we have

$$v(j,1) - v(j,0)$$
$$\geq c[(j,1), a_0] - \mathsf{E}c[(j,0), a_2^{X_0}] - g\{\bar{\tau}[(j,1), a_0] - \mathsf{E}\bar{\tau}[(j,0), a_2^{X_0}]\}$$
$$+(1-\theta)\sum_{i \leq j} p_{(j,1),(i,0)}(a_0)[v(i+1,0) - v(i,0)]$$

$$\geq \; C_I + (1-\theta)C_{P2} - (1-\theta)R + (1-\theta)\sum_{i\leq j} p_{(j,1),(i,0)}(a_0) i h \mathsf{E}[Y]$$

$$\geq \; C_I + (1-\theta)C_{P2} - (1-\theta)R + (1-\theta)h\mathsf{E}[Y]\{j - \mathsf{E}[N(X_0)]\}$$

$$= \; C_I + (1-\theta)C_{P2} - (1-\theta)R + (1-\theta)(jh\mathsf{E}[Y] - h\mathsf{E}[X_0])$$

$$= \; \Delta(j).$$

Here the second inequality makes use of Lemma 5.11 and the facts that

$$c[(j,1),a_0] - \mathsf{E}c[(j,0),a_2^{X_0}] = C_I + (1-\theta)(C_{P_2} - R)$$

and

$$\bar{\tau}[(j,1),a_0] = \mathsf{E}\bar{\tau}[(j,0),a_2^{X_0}] = \mathsf{E}[X_0].$$

The third inequality takes into account that the number of departures in M_2 during X_0 time units is at most $N(X_0)$. The first equality is due to the relation that $\mathsf{E}[N(X_0)] = \mathsf{E}[X_0]/\mathsf{E}[Y]$. □

Theorem 5.13 The optimal policy forgoes the processing action a_1 in all states $(i,1)$ and $(i,0)$ with $i \geq i^*$, where

$$i^* \; := \; \min\{i : \mathsf{E}[i - N(X_1)]^+$$
$$\geq \frac{(1-\theta)(R - C_{P2}) - C_{P1} - C_I + (1-\theta)h\mathsf{E}[X_0]}{(1-\theta)h\mathsf{E}[Y]}\}.$$

More specifically, in such states, the idling action is always preferred to the processing action.

Proof. We show that in any state $(i,1)$ or $(i,0)$, with $i \geq i^*$, action $a_2^{X_1}$ is at least as good as action a_1. Here $a_2^{X_1}$ denotes the action a_2 with an idling time $T = X_1$, a random sample from the distribution of X_1, the processing time at M_1. From Theorem 5.9, it suffices to show that in state $(i,1)$ the following relation holds:

$$c[(i,1),a_1] - g\bar{\tau}[(i,1),a_1] + \sum_j p_{(i,1),(j,1)}(a_1)v(j,1) \qquad (5.21)$$

$$\geq \; \mathsf{E}c[(i,1),a_2^{X_1}] - g\mathsf{E}\bar{\tau}[(i,1),a_2^{X_1}] + \sum_j p_{(i,1),(j,0)}(a_2^{X_1})v(j,0).$$

Note that

$$\bar{\tau}[(i,1),a_1] = \mathsf{E}\bar{\tau}[(i,1),a_2^{X_1}] = \mathsf{E}[X_1],$$
$$c[(i,1),a_1] - \mathsf{E}c[(i,1),a_2^{X_1}] = C_{P1},$$

and

$$p_{(i,1),(j,1)}(a_1) = p_{(i,1),(j,0)}(a_2^{X_1}).$$

Hence, (5.21) is equivalent to

$$\sum_j p_{(i,1),(j,1)}(a_1)[v(j,1) - v(j,0)] \geq -C_{P1}.$$

For this to hold, from Lemma 5.12, it suffices to have

$$C_I + (1-\theta)(C_{P2} - R) - (1-\theta)hE[X_0]$$
$$+(1-\theta)hE[Y]\sum_j jp_{(i,1),(j,1)}(a_1) \geq -C_{P1}. \qquad (5.22)$$

From the transition probabilities in §5.1, we have

$$\sum_j jp_{(i,1),(j,1)}(a_1) = \sum_{j=1}^{i+1} jP[N(X_1) = i + 1 - j]$$

$$= \sum_{k=0}^{i}(i + 1 - k)P[N(X_1) = k]$$

$$= E[i + 1 - N(X_1)]^+.$$

Hence, (5.22) is equivalent to

$$E[i + 1 - N(X_1)]^+$$
$$\geq \frac{(1-\theta)(R - C_{P2}) - C_{P1} - C_I + hE[X_0]}{(1-\theta)hE[Y]}. \qquad (5.23)$$

Because the left-hand side is increasing in i, the inequality holds for all $i \geq i^* - 1$; in particular, it holds for $i \geq i^*$.

Similarly, in state $(i, 0)$, $a_2^{X_1}$ is preferred if

$$c[(i,0), a_1] - g\bar{\tau}[(i,0), a_1] + \sum_j p_{(i,0),(j,1)}(a_1)v(j,1)$$

$$\geq Ec[(i,0), a_2^{X_1}] - gE\bar{\tau}[(i,0), a_2^{X_1}]$$
$$+ \sum_j p_{(i,0),(j,0)}(a_2^{X_1})v(j,0),$$

which, analogous to (5.21), is equivalent to

$$\sum_j p_{(i,0),(j,1)}(a_1)[v(j,1) - v(j,0)] \geq -c_{P1},$$

which, in turn, is implied by [cf. (5.23)]

$$E[i - N(X_1)]^+ \geq \frac{(1-\theta)(R - C_{P2}) - C_{P1} - C_I + (1-\theta)hE[X_0]}{(1-\theta)hE[Y]}.$$

Hence, it suffices to have $i \geq i^*$. $\square$

Theorem 5.13 implies the following fact.

Proposition 5.14 The maximum required buffer capacity between the two stages is $i^* + 1$, where i^* is the threshold value specified in Theorem 5.13. In other words, under an optimal policy, the number of jobs in the buffer will never exceed $i^* + 1$.

Proof. A quick examination of the state transition probabilities in §5.1.1 indicates that no state transition can increase the buffer content by more than one unit. Hence, we only need to examine the possible actions and transitions in states with $i = i^*$ and $i^* + 1$.

In state $(i^*, 0)$, the only action is to stay idling (which is preferred to processing, following Theorem 5.13); hence the buffer content cannot increase. In state $(i^*, 1)$, from the proof of Theorem 5.9 (in particular (5.23)), again the processing action is ruled out; and the worst-case scenario, in terms of the highest buffer content, is to transit into state $(i^* + 1, 0)$ via a (successful) inspection (and no job completion at M_2 during the inspection). Once in state $(i^* + 1, 0)$, the only action is to remain idling.

It is also clear from the preceding discussion that the state $(i^* + 1, 1)$ can never be reached. So in summary, the buffer content will never exceed $i^* + 1$. □

5.4 Notes

Buzacott and Shanthikumar [11, 12], chapters 4 and 5 in particular, provide details for the modeling and performance analysis of make-to-stock queues, in both isolation and in tandem. More general than the base-stock control are the "kanban" control (e.g., Cheng and Yao [22], Glasserman and Yao [39, 42]), and PAC (production authorization card) control schemes (chapter 10 of [11]). A common feature of these control mechanisms is that they all lead to some form of buffer capacity limit at each stage. This limit is important, not only because it bounds the work-in-process inventory (and, consequently, the cycle time of jobs), but also because it is central to the physical design of the system (for example, in terms of setting storage space between two consecutive stages).

In Chen, Yang, and Yao [17], a similar control problem is studied; the first stage serves two types of traffic: type 1 leaves the system after service, whereas type 2 jobs continue into a second stage. Among other things, the optimal switching between serving the two types at the first stage is considered. The control actions are serving type 1, serving type 2 or staying idle. This is quite similar to our problem of controlling the production, inspection, or idling processes at stage 1. The difference, however, is that here production and inspection are performed on the same job, and only defective jobs, identified via inspection, leave the system—corresponding to type 1 jobs in [17]; all other jobs continue into stage 2.

A more general class of optimal switching control (as well as rate control) problems in queueing networks is studied in Glasserman and Yao [41] (also see [40], chapter 6). Whereas both [17] and [41] consider Markov decision programs with discounted objectives, here we study a semi-Markov decision program with a long-run average objective. For general results on SMDP with long-run average objectives, see Federgruen and Tijms [34] and the references there.

This chapter is based on Yao and Zheng [109]. There are several possible extensions of the results here. First, further characterization of the optimal policy is possible. For example, in addition to the upper threshold i^* in Theorem 5.13, it is possible to establish a lower threshold, i.e., below the threshold processing is preferred to idling. It is also possible to characterize similar scenarios in which inspection is or is not preferred. All of these, however, seem to require additional assumptions on relations among the given cost data and/or processing and inspection times (i.e., X_0, X_1, and Y). (In addition, those properties of the one-step cost function—see the remark following the proof of Lemma 5.1—not yet used here may also play a role.) In view of the model in [17], the need for additional assumptions is perhaps no surprise: a total of six scenarios are identified in [17], in terms of the relations among given data, and the optimal control takes a different form in each of the six scenarios. It would also be worthwhile to derive heuristic policies for the control problem of this chapter that are similar to some of those in [17].

A second possibility is to allow a *random* defective rate, known only in terms of distribution, and instead of deciding whether to inspect immediately after every item is completed at the first stage, the inspection decision can take place once in a while when there are a few items in the buffer. Furthermore, the inspection can be carried out in a sequential fashion, i.e., as defects are identified, knowledge about the conditional distribution of the defective rate is updated, and the decision of whether to continue inspection is made accordingly. This is similar to the sequential inspection problem in Chapter 3. The additional difficulty here, of course, lies in the fact that this optimal stopping problem is interweaved with the optimal switching problem (among processing, inspection, and idling).

6

Sequential Inspection Under Capacity Constraints

We now combine the inspection process of Chapter 3 with the tandem system configuration of Chapter 5: there is an inspection facility located between two production stages; it takes in batches of jobs from an upstream stage and feeds the batches, after inspection, into a downstream stage. A motivating example is the production-inspection process at a semiconductor wafer fabrication facility ("fab"), where inspection corresponds to the so-called 'wafer probe', which inspects the chips on each wafer. The upstream stage is wafer fabrication, which sends completed wafers to the probe stage, where defective chips are identified. The wafers then go into a stage further downstream, where the defective chips are discarded or sent to rework, and the good chips go through final testing and packaging.

The coordination among the capacities of the inspection facility and its upstream and downstream partners translates into two constraints on the inspection policy: the expected number of units inspected cannot exceed a certain upper limit—so that input from the upstream can be handled without undue delay, and the expected number of defective units identified must exceed a certain lower limit—so that the output from the inspection will not overpower the capacity at the downstream (which does not have to deal with identified defective items).

Here we demonstrate that the model in Chapter 3 can be extended to a *constrained* MDP, with two constraints reflecting the capacity limits. We show that the optimality of the sequential policy still holds, with appropriate *randomization* at the thresholds.

In §6.1, we introduce the constrained model and explore the dynamic programming recursion of the MDP. In §6.2, we prove that the sequential threshold policy of the unconstrained model remains optimal for the constrained MDP, with possible randomization at the thresholds. In §6.3, we further show that the optimal threshold policy can be obtained by solving a certain LP and randomization is needed at no more than two thresholds. In §6.4, we illustrate the application of the model in the context of semiconductor wafer fabrication mentioned earlier.

6.1   Capacity Constraints

Here we study the same problem as in Chapter 3 of finding a policy π that minimizes the overall cost associated with a batch of N items, but with the addition of two constraints as follows. Let I^π and D^π be, respectively, the number of inspected units and the number of defectives identified under policy π. Let u and ℓ be two given constants, reflecting the capacity associated with upstream and downstream production stages. Then, the two constraints are:

$$\mathsf{E}[I^\pi] \leq u \quad \text{and} \quad \mathsf{E}[D^\pi] \geq \ell. \tag{6.1}$$

Note, in particular, the second inequality means that, for a given input flow, we should remove enough defective units to reduce the workload of the downstream stage. (Refer to §6.4 for more motivation on u and ℓ.)

Observe that the total expected cost under policy π, which stops after inspecting I^π units and having identified a total of D^π defectives, is

$$V_0^\pi(0) := c_i \mathsf{E}[I^\pi] + c_r \mathsf{E}[D^\pi] + \mathsf{E}[\phi(I^\pi, \Theta)], \tag{6.2}$$

where $\phi(\cdot)$ is defined in (3.1). Hence, our new problem is:

$$\min_\pi \quad V_0^\pi(0) \tag{6.3}$$
$$\text{subject to} \quad \mathsf{E}[I^\pi] \leq u, \quad \mathsf{E}[D^\pi] \geq \ell.$$

We want to show that in the presence of the constraints, the sequential optimal policy of Chapter 3, with minor modification, remains optimal. In particular, the form of the optimal policy is still characterized by a sequence of thresholds $\{d_n^*; n \leq N - 1\}$, with possible randomization at the thresholds. To this end, we need to explore deeper the structure of the dynamic programming recursion.

Let (n, d) denote the state of decision: n units have been inspected and d units are found defective. The state space is

$$\mathcal{S} := \{(n, d) : d \leq n; \, n = 0, 1, \ldots, N - 1\}. \tag{6.4}$$

A policy π is characterized by the action it prescribes in each state: to continue or to stop inspection, denoted a_1 and a_0, respectively. In general (taking into account the need for randomization), π takes the following form: for $i = 0, 1$, in state (n, d) choose action a_i with probability $p(n, d; a_i)$, where $p(n, d; a_i) \in [0, 1]$ and $p(n, d; a_0) + p(n, d; a_1) = 1$. Later, when we write $\pi(n, d) = a_i$, for $i = 0$ or 1, we mean with probability 1, the prescribed action by π in state (n, d) is a_i.

Whenever necessary, we will use the superscript π to index quantities associated with the policy π. Let

$$P^\pi(n, d) := \mathsf{P}[I^\pi \geq n, D(n) = d]$$

be the probability of the occurrence of the state (n, d). We have

$$
\begin{aligned}
& P^\pi(n, d) \\
= \ & \sum_{i=0,1} \mathsf{P}[I^\pi \geq n, D(n) = d, D(n-1) = d - i] \\
= \ & \sum_{i=0,1} \mathsf{P}[I^\pi \geq n, D(n-1) = d - i]\mathsf{P}[D(n) = d | D(n-1) = d - i] \\
= \ & \sum_{i=0,1} \mathsf{P}[I^\pi \geq n - 1, D(n-1) = d - i, \pi(n-1, d-i) = a_1] \\
& \qquad \cdot \mathsf{P}[D(n) = d | D(n-1) = d - i].
\end{aligned}
$$

Because

$$
\begin{aligned}
& \mathsf{P}[I^\pi \geq n - 1, D(n-1) = d - i, \pi(n-1, d-i) = a_1] \\
= \ & P^\pi(n-1, d-i)p^\pi(n-1, d-i; a_1),
\end{aligned}
$$

we have, for $1 < n \leq N - 1$ and $d \leq n$,

$$
\begin{aligned}
P^\pi(n, d) \ = \ & \sum_{i=0,1} P^\pi(n-1, d-i)p(n-1, d-i; a_1) \\
& \qquad \cdot \mathsf{P}[D(n) = d | D(n-1) = d - i] \quad (6.5)
\end{aligned}
$$

and

$$P^\pi(1, i) = p(0, 0; a_1)\mathsf{P}[D(1) = i], \quad i = 0, 1. \quad (6.6)$$

Hence, the probabilities $P^\pi(n, d)$ can be recursively computed from (6.5) and (6.6). Note that in the recursions, for $i = 0, 1$,

$$\mathsf{P}[D(1) = 1] = \mathsf{E}[\Theta], \quad \mathsf{P}[D(1) = 0] = 1 - \mathsf{E}[\Theta],$$

$$
\begin{aligned}
& \mathsf{P}[D(n) = d | D(n-1) = d - 1] \\
= \ & \mathsf{E}[\Theta | D(n-1) = d - 1] \\
= \ & \mathsf{E}[\Theta^d(1 - \Theta)^{n-d}]/\mathsf{E}[\Theta^{d-1}(1 - \Theta)^{n-1-d+1}], \quad (6.7)
\end{aligned}
$$

and

$$P[D(n) = d | D(n-1) = d]$$
$$= 1 - E[\Theta | D(n-1) = d]$$
$$= E[\Theta^d (1-\Theta)^{n-d}] / E[\Theta^d (1-\Theta)^{n-1-d}]. \tag{6.8}$$

Here (6.7) and (6.8) are from (3.8).

Now suppose we have inspected a total of n items. Then the cost difference between the two alternatives—stopping inspection after inspecting one more unit versus stopping immediately—is as follows:

$$\Delta(n) \quad := \quad c_i + c_r E[\Theta | D(n)] + E[\phi(n+1, \Theta) | D(n)]$$
$$-E[\phi(n, \Theta) | D(n)], \quad n \leq N-1. \tag{6.9}$$

Note that from Lemma 3.10,

$$E[\Delta(n) | D(n) = d] - c_i$$
$$= c_r E[\Theta_n(d)] + E[\phi(n+1, \Theta_n(d))] - E[\phi(n, \Theta_n(d))]$$

is decreasing in d and increasing in n.

For a given policy π, define

$$V^\pi \quad := \quad \sum_{n=0}^{N-1} \sum_{d \leq n} E[\Delta(n) | D(n) = d] P^\pi(n, d) p(n, d; a_1)$$

$$= \quad \sum_{n=0}^{N-1} E[\Delta(n)] P[I^\pi > n]$$

$$= \quad E[\Delta(0) + \Delta(1) + \cdots + \Delta(I^\pi - 1)]. \tag{6.10}$$

Because

$$E[D^\pi] = E[\Theta | D(0)] + E[\Theta | D(1)] + \cdots + E[\Theta | D(I^\pi - 1)],$$

from (6.9) and (6.10), we have

$$V^\pi \quad = \quad c_i E[I^\pi] + c_r E[D^\pi] + E\{E[\phi(I^\pi, \Theta) | D(I^\pi - 1)]\} - E[\phi(0, \Theta)]$$
$$= \quad c_i E[I^\pi] + c_r E[D^\pi] + E[\phi(I^\pi, \Theta)] - E[\phi(0, \Theta)]$$
$$= \quad V_0^\pi(0) - E[\phi(0, \Theta)]. \tag{6.11}$$

Because $E[\phi(0, \Theta)]$ is independent of the policy, we can conclude that the original problem of minimizing $V_0^\pi(0)$ is equivalent to minimizing V^π.

In what follows, we will discuss the problem of finding a policy π that minimizes V^π and satisfies the two constraints in (6.1), i.e.,

$$\min_\pi \quad V^\pi \tag{6.12}$$

$$\text{subject to} \quad E[I^\pi] \leq u, \quad E[D^\pi] \geq \ell.$$

Remark 6.1 There is an intuitive interpretation of V^π. Because

$$-\mathsf{E}[\Delta(n)|D(n) = d]$$

can be viewed as the cost reduction obtained from inspecting an item in state (n, d), $-V^\pi$ is the total cost reduction under policy π. Hence, the original problem can also be viewed as finding a policy π that maximizes $-V^\pi$, the total cost reduction.

6.2   Optimality of the Threshold Policy

We start with a precise statement about what constitutes a threshold policy in the setting of our constrained MDP.

Definition 6.2 A policy π is said to be of *threshold type*, if, for $n = 1, 2, \cdots, N - 1$, there exist integers d_n^* such that $\pi(n, d) = a_1$ for $d > d_n^*$ and $\pi(n, d) = a_0$ for $d < d_n^*$.

Remark 6.3 Definition 6.2 does not specify what actions to take at the threshold values: $d = d_n^*$, $n = 0, 1, ..., N - 1$. In general, the actions can be randomized: at d_n^* take a_1 with probability $q^\pi(d_n^*)$, and take a_0 with probability $1 - q^\pi(d_n^*)$, where $q^\pi(d_n^*)$ is a parameter (specified by the policy π). This randomization at the threshold point is necessary in particular for cases when the constraints become tight. More discussions about the randomization are presented in §6.3.

Remark 6.4 Note that Definition 6.2 is equivalent to the following alternative statement:

- a policy π is said to be of threshold type, if, for each $n = 1, 2, \cdots, N - 1$, $\mathsf{P}[\pi(n, d) = a_1] > 0$ implies $\mathsf{P}[\pi(n, d') = a_1] = 1$ for any $d' : n \ge d' > d$.

Clearly, this is implied by Definition 6.2: for each n, $\mathsf{P}[\pi(n, d) = a_1] > 0$ can only hold for $d : n \ge d \ge d_n^*$, and hence $\mathsf{P}[\pi(n, d') = a_1] = 1$ as $d' > d^*$. On the other hand, Definition 6.2 can be recovered by defined d_n^* as:

$$d_n^* := \min\{d : \mathsf{P}[\pi(n, d) = a_1] > 0\}$$

for each n.

Lemma 6.5 Let Θ and Θ' be the defective ratios of two batches (of the same size). Let $D(n)$ and $D'(n)$ be the number of defective units in n items, respectively for the two batches. Then, $\Theta \le_{lr} \Theta'$ implies

(i) $D(n) \le_{lr} D'(n)$ and

(ii) $P[D(n+1) \geq b|D(n) \geq a] \leq P[D'(n+1) \geq b|D'(n) \geq a]$, for any integers a and b satisfying $0 \leq a \leq n$, $0 \leq b \leq n+1$.

Proof. (i) The desired likelihood ratio ordering is equivalent to

$$P[D'(n) = k]/P[D(n) = k] = E[(\Theta')^k(1-\Theta')^{n-k}]/E[\Theta^k(1-\Theta)^{n-k}]$$

increasing in k, for $k \leq n$. Hence, it suffices to show

$$E[(\Theta')^k(1-\Theta')^{n-k}]E[\Theta^{k+1}(1-\Theta)^{n-k-1}]$$
$$\leq E[(\Theta')^{k+1}(1-\Theta')^{n-k-1}]E[\Theta^k(1-\Theta)^{n-k}].$$

Let $f(\cdot)$ and $g(\cdot)$ be the density functions of Θ' and Θ. Then, the preceding can be written explicitly as

$$\int_y \int_x [x^{k+1}(1-x)^{n-k-1}y^k(1-y)^{n-k}$$
$$-x^k(1-x)^{n-k}y^{k+1}(1-y)^{n-k-1}]f(x)g(y)dxdy$$
$$\geq 0.$$

Next, separate the integral over x into two parts: $\int_{x \geq y}$ and $\int_{x \leq y}$, and interchange x and y in the second part. This then becomes

$$\int_y \int_{x \geq y} [x^{k+1}(1-x)^{n-k-1}y^k(1-y)^{n-k}$$
$$-x^k(1-x)^{n-k}y^{k+1}(1-y)^{n-k-1}][f(x)g(y) - f(y)g(x)]dxdy$$
$$\geq 0.$$

This clearly holds, because the quantities in the first pair of brackets are nonnegative, due to $x \geq y$; and those in the second pair are also nonnegative, due to $\Theta \leq_{\mathrm{lr}} \Theta'$. Hence, the desired result follows.

(ii) If $a \geq b$, the inequality is trivial because both sides equal 1. In what follows we assume $a < b$.

First observe that $D(n+1) \geq b$ implies $D(n) \geq a$, because

$$D(n) \geq D(n+1) - 1$$

and $b \geq a + 1$. Hence,

$$P[D(n+1) \geq b|D(n) \geq a] = P[D(n+1) \geq b]/P[D(n) \geq a],$$

and we want to establish the following inequality:

$$P[D(n+1) \geq b]P[D'(n) \geq a] \leq P[D'(n+1) \geq b]P[D(n) \geq a]. \qquad (6.13)$$

Note that

$$P[D(n) \geq a] = \sum_{k=a}^{n} \binom{n}{k} E[\Theta^k(1-\Theta)^{n-k}]$$

and

$$P[D(n+1) \geq b] = \sum_{k=b}^{n+1} \binom{n+1}{k} E[\Theta^k (1-\Theta)^{n+1-k}]$$

$$= \sum_{k=b-1}^{n} \binom{n+1}{k+1} E[\Theta^{k+1} (1-\Theta)^{n-k}].$$

Hence, for (6.13) to hold, it suffices to have

$$E[\Theta^{j+1}(1-\Theta)^{n-j}] E[(\Theta')^k (1-\Theta')^{n-k}]$$
$$\leq E[(\Theta')^{k+1}(1-\Theta')^{n-k}] E[\Theta^j (1-\Theta)^{n-j}],$$

for any $j, k \geq b-1$; and

$$E[\Theta^{j+1}(1-\Theta)^{n-j}] E[(\Theta')^i (1-\Theta')^{n-i}]$$
$$\leq E[(\Theta')^{j+1}(1-\Theta')^{n-j}] E[\Theta^i (1-\Theta)^{n-i}],$$

for $i \leq a \leq b-1 \leq j$. But these can be verified following the same argument as in (i). □

Lemma 6.6 Consider two batches with the same size but different defective ratios, $\Theta_1 \leq_{lr} \Theta_2$. Suppose π is a threshold policy for inspecting the first batch. Then there exists a policy π' (not necessarily of threshold type) for inspecting the second batch, such that

$$I_1^\pi =_{st} I_2^{\pi'}, \qquad D_1^\pi \leq_{st} D_2^{\pi'}$$

(where $=_{st}$ denotes equal in distribution, and the subscripts 1 and 2 refer to the two batches); and hence

$$E[I_1^\pi] = E[I_2^{\pi'}], \qquad E[D_1^\pi] \leq E[D_2^{\pi'}].$$

Furthermore, π' achieves a larger cost reduction in the second batch, i.e., $V_1^\pi \geq V_2^{\pi'}$.

Proof. For each $n \leq N-1$, let $D(n)$ and $D'(n)$ denote the number of defective items identified from the two batches identified through inspecting n items in each batch. Then, we have $D'(n) \geq_{lr} D(n)$, following Lemma 6.5 (i). Suppose d_n^* for $n = 1, 2, \cdots, N-1$ are the corresponding thresholds of π. Denote

$$p_1 = P[D(1) \geq d_1^*], \qquad p_1' = P[D'(1) \geq d_1^*],$$

$$q_0 = P[\pi(0, 0) = a_1], \qquad q_1 = P[\pi(1, D(1)) = a_1 | D(1) \geq d_1^*];$$

and for $2 \leq n \leq N-1$, denote

$$p_n = P[D(n) \geq d_n^* | D(n-1) \geq d_{n-1}^*],$$

$$p'_n = \mathsf{P}[D(n)' \geq d^*_n | D'(n-1) \geq d^*_{n-1}],$$

and

$$q_n = \mathsf{P}[\pi(n, D(n)) = a_1 | D(1) \geq d^*_1, D(2) \geq d^*_2, ..., D(n) \geq d^*_n].$$

Then $p'_n \geq p_n$, for all $n = 1, ..., N - 1$, follows from Lemma 6.5 (ii).

Now, given the policy π, we construct a new policy π' as follows: let

$$\pi'(n, d) = \pi(n, d) = a_0, \quad \text{for} \quad d < d^*_n;$$

and for $d \geq d^*_n$,

$$\pi'(n, d) = \left\{ \begin{array}{ll} a_1 & \text{w.p.} \quad q_n p_n / p'_n, \\ a_0 & \text{w.p.} \quad 1 - q_n p_n / p'_n. \end{array} \right.$$

Observe the following:

$$\begin{aligned}
&\mathsf{P}[I^{\pi'}_2 > n] \\
=\ & q_0 \mathsf{P}[D'(i) \geq d^*_i, \pi'(i, D'(i)) = a_1; 1 \leq i \leq n] \\
=\ & q_0 \frac{q_1 p_1}{p'_1} \mathsf{P}[D'(1) \geq d^*_1] \frac{q_2 p_2}{p'_2} \mathsf{P}[D'(2) \geq d^*_2 | D'(1) \geq d^*_1] \\
&\cdots \frac{q_n p_n}{p'_n} \mathsf{P}[D'(n) \geq d^*_n | D'(n-1) \geq d^*_{n-1}] \\
=\ & q_0 (p_1 q_1)(p_2 q_2) \cdots (p_n q_n) \\
=\ & \mathsf{P}[I^{\pi}_1 > n].
\end{aligned}$$

Here the first equality follows from the fact that $[I^{\pi'}_2 > n]$ means a_1 is used for all $i = 0, 1, \cdots, n$, and a_1 is used at i for $i \geq 1$ only if $D'(i) \geq d^*_i$ and $\pi'(i, D'(i)) = a_1$. Hence, we have

$$I^{\pi}_1 =_{\text{st}} I^{\pi'}_2.$$

To show the other results, given n, conditioning on $I^{\pi}_1 = I^{\pi'}_2 = n$, and through coupling, we have $D(n) \leq D'(n)$ (almost surely). Hence, $D^{\pi}_1 \leq_{\text{st}} D^{\pi'}_2$. Furthermore, $V^{\pi}_1 \geq V^{\pi'}_2$ because $\mathsf{E}[\Delta(n)] \geq \mathsf{E}[\Delta'(n)]$, following Lemma 3.10.□

Now we are ready to prove the main result.

Theorem 6.7 For the constrained MDP as characterized by the objective function in (6.11) and the constraints in (6.1), if there exists an optimal policy (minimizing V^{π}), then there must exist an optimal policy of the threshold type specified in Definition 6.2.

Proof. Suppose π is an optimal policy that is *not* of threshold type. In particular, suppose it violates Definition 6.2. That is, for some n,

$$\mathsf{P}[\pi(n, d) = a_1] = p > 0 \quad \text{whereas} \quad \mathsf{P}[\pi(n, d') = a_1] = p' < 1, \quad (6.14)$$

for some pair (d, d') with $d' > d$. We want to show, through induction on n, that there exists another policy π', that corrects the violation (and is hence of threshold type), while satisfying

$$I^{\pi'} =_{\text{st}} I^{\pi}, \quad \mathsf{E}[D^{\pi'}] \geq \mathsf{E}[D^{\pi}], \quad \text{and} \quad V^{\pi'} \leq V^{\pi}. \tag{6.15}$$

That is, π' is a threshold policy that is feasible and performs at least as well as π (in terms of the objective value).

(I) The Initial Step. Consider $n = N - 1$. Construct π' as follows. Write $P := P^{\pi}(N - 1, d)$ and $P' := P^{\pi}(N - 1, d')$. There are two cases:
Case (i): $P'(1 - p') \leq Pp$. Let

$$\pi'(s) = \pi(s), \qquad s \neq (N - 1, d), (N - 1, d');$$

let

$$\pi'(N - 1, d') = a_1$$

and

$$\pi'(N - 1, d) = \begin{cases} a_1 & \text{w.p.} \quad p - (1 - p')P'/P, \\ a_0 & \text{w.p.} \quad 1 - p + (1 - p')P'/P. \end{cases}$$

That is, under policy π', the probability of choosing action a_1 is increased by an amount $1 - p'$ at $(N - 1, d')$ (i.e., from p' to 1, see (6.14)), and decreased by an amount $(1 - p')P'/P$ at $(N - 1, d)$. To argue that $I^{\pi'} =_{\text{st}} I^{\pi}$, it suffices to show $\mathsf{P}[I^{\pi'} = k] = \mathsf{P}[I^{\pi} = k]$ for $0 \leq k \leq N$.

Clearly, $\mathsf{P}[I^{\pi} = k]$ only depends on $\pi(i, d_i)$ for $i \leq k$ and $d_i \leq i$. Because $\pi(i, d_i) = \pi'(i, d_i)$ for any $i < N - 1$,

$$\mathsf{P}[I^{\pi'} = k] = \mathsf{P}[I^{\pi} = k] \qquad \text{for } 0 \leq k < N - 1.$$

On the other hand,

$$p^{\pi'}(N - 1, d'; a_1) = p^{\pi}(N - 1, d'; a_1) + (1 - p'),$$
$$p^{\pi'}(N - 1, d; a_1) = p^{\pi}(N - 1, d; a_1) - (1 - p')P'/P.$$

Furthermore, because π' follows the same actions as π except at $n = N - 1$, we have

$$P^{\pi'}(N - 1, d_{N-1}) = P^{\pi}(N - 1, d_{N-1})$$

for any $d_{N-1} \leq N - 1$. Hence, we have

$$\begin{aligned}
&\mathsf{P}[I^{\pi'} = N] \\
={} & \mathsf{P}[I^{\pi} = N] + P^{\pi'}(N - 1, d')(1 - p') - P^{\pi'}(N - 1, d)(1 - p')P'/P \\
={} & \mathsf{P}[I^{\pi} = N] + P^{\pi}(N - 1, d')(1 - p') - P^{\pi}(N - 1, d)(1 - p')P'/P \\
={} & \mathsf{P}[I^{\pi} = N].
\end{aligned}$$

In addition,

$$P[I^{\pi'} = N - 1] = P[I^{\pi} = N - 1]$$

follows from

$$\sum_{k=0}^{N} P[I^{\pi'} = k] = \sum_{k=0}^{N} P[I^{\pi} = k] = 1.$$

Hence, $I^{\pi'} =_{st} I^{\pi}$.

Similarly, by conditioning on the occurrence of (n, d) and (n, d') we have

$$
\begin{aligned}
E[D^{\pi'}] &= E[D^{\pi}] - P[(1 - p')P'/P]E[\Theta|d] + (1 - p')P'E[\Theta|d'] \\
&= E[D^{\pi}] + (1 - p')P'(E[\Theta|d'] - E[\Theta|d]) \\
&\geq E[D^{\pi}],
\end{aligned}
\tag{6.16}
$$

where the inequality follows from the fact that $d' \geq d$ implies $E[\Theta|d'] > E[\Theta|d]$. (Refer to Lemma 3.9.) Define

$$\Delta(n, d) := E[c_r\Theta + \phi(n + 1, \Theta) - \phi(n, \Theta)|D(n) = d].$$

Then $\Delta(n, d)$ is decreasing in d, following Lemma 3.10 . Replacing $E[\Theta|d']$ and $E[\Theta|d]$ by $\Delta(N - 1, d')$ and $\Delta(N - 1, d)$, we have,

$$
\begin{aligned}
&E[\Delta'(N - 1)] \\
= \ &E[\Delta(N - 1)] - (1 - p')P'[\Delta(N - 1, d) - \Delta(N - 1, d')] \\
\leq \ &E[\Delta(N - 1)].
\end{aligned}
\tag{6.17}
$$

Hence, $V^{\pi'} \leq V^{\pi}$.

Case (ii): $P'(1 - p') > Pp$. Let

$$\pi'(s) = \pi(s), \qquad s \neq (N - 1, d), (N - 1, d');$$

let

$$\pi'(N - 1, d) = a_0$$

and

$$\pi'(n, d') = \begin{cases} a_1 & \text{w.p.} \quad p' + pP/P', \\ a_0 & \text{w.p.} \quad (1 - p') - pP/P'. \end{cases}$$

That is, policy π' reduces the probability of choosing action a_1 in $(N-1, d)$ by an amount p (to 0) and increases the same probability in $(N - 1, d')$ by an amount pP/P'. It is easy to verify that the three relations in Case (i) remain intact.

Hence, in either case, for $n = N - 1$, we can correct the violation of the threshold property at any pair (d, d') by repeatedly applying the outlined procedure. The resulting policy π' will then satisfy the threshold property (at n). Also, the new policy still satisfies the constraints in (6.1), and it can only improve the objective function.

(II) The Induction Step. Now, as induction hypothesis, suppose for any policy π that violates the threshold property for $n \geq m$ can be replaced by another policy π' that corrects all the violations, and satisfies (6.15).

Now, suppose π violates the threshold property at $n = m - 1$. We can carry out the corrections in exactly the same manner as in the case of $n = N - 1$, and denote the resulting policy as π^1. However, unlike the case of $n = N - 1$, we also have to specify the actions of π^1 for $n \geq m$. Suppose we simply let π^1 follow the same actions as π for $n \geq m$ (as well as for $n \leq m-2$). Then we do not have $I^{\pi^1} =_{\mathrm{st}} I^\pi$. This is because starting from the next inspection (i.e., $n = m$), the two policies will have to face two (partial) batches of different quality. Specifically (using the superscript 1 to denote quantities that relate to π^1), because π^1 has more chance to identify a defective unit at $n = m - 1$ (through coupling), we have $D_{m-1} \leq D^1_{m-1}$, and hence

$$[\Theta|D_{m-1}] \leq_{\mathrm{lr}} [\Theta|D^1_{m-1}]$$

(refer to Lemma 3.9). Because π^1 is constructed as always taking the same actions as π, for $n \geq m$, π^1 will end up with inspecting more items.

However, we can make use of Lemma 6.6. (Note that the relation

$$[\Theta|D_{m-1}] \leq_{\mathrm{lr}} [\Theta|D^1_{m-1}]$$

puts us right in the setting of the lemma, for two batches, each of size $N - m$.) Specifically, we can construct another policy, π^2, which follows the same actions as π^1 up to $n = m - 1$. But for $n \geq m$, it modifies the actions of π in the same way as π' of Lemma 6.6, so that the number of items inspected under it will follow the same distribution as that of π (while it will do better in the number of defective units identified). Nevertheless, π^2 could still violate the threshold property, for $n \geq m$. But then we can make use of the induction hypothesis to construct yet another *threshold* policy that matches π^2 in terms of the number of items inspected (and it still does better in the expected number of defective units identified). This is the threshold policy we want.

Note that the final policy obtained earlier, denoted π', pieces together the actions of three policies: those of π for $n \leq m - 2$, those of π^1 for $n = m - 1$ and those of π^2 for $n \geq m$. What remains is to establish that $V^{\pi'} \leq V^\pi$. Because $I^{\pi'}$ and I^π are equal in distribution, denote both by I, and condition on I. If $I \leq m - 2$, the two policies yield the same objective value. If $I \geq m - 1$, then $V^{\pi'}$ is smaller, because from $n = m - 1$ and beyond, π' follows first π^1 and then π^2, both incurring smaller $\mathsf{E}[\Delta(n)]$ values than π (as evident from the case of $n = N - 1$ and the proof of Lemma 6.6). $\square$

6.3 Further Characterization of the Optimal Policy

To completely derive the optimal policy characterized in the last section, we still need to find the optimal thresholds, and the optimal randomization (if any) at each threshold. We will show that this can be accomplished by solving a linear program, and that randomization is needed at no more than two threshold values.

We first model our problem as an infinite-horizon constrained MDP that minimizes a long-run average-cost objective. Let i^t and a^t denote the system state and the control action chosen at time t. Recall that, the action space contains only two actions: $a^t \in \{a_0, a_1\}$ for all t (i.e., either stop or continue inspection). For the state space, however, we enlarge the space $\mathcal{S}$ in (6.4) by adding to it n "pseudo states" as follows:

$$\mathcal{S}' := \{s(n) : 0 \leq n \leq N - 1\}.$$

These pseudo states will be visited once the action a_0 (stopping inspection) is taken. Specifically, transition probability function $p_{i,j}(a)$ is as follows: When $(n, d) \in \mathcal{S}$, for $n = N - 1$,

$$p_{(N-1,d),(0,0)}(a_0) = p_{(N-1,d),(0,0)}(a_1) = 1;$$

while for $n < N - 1$,

$$p_{(n,d),(n+1,d+1)}(a_1) = \mathsf{E}[\Theta | D(n) = d],$$

$$p_{(n,d),(n+1,d)}(a_1) = 1 - \mathsf{E}[\Theta | D(n) = d],$$

and

$$p_{(n,d),s(n+1)}(a_0) = 1.$$

Furthermore, for $s(n) \in \mathcal{S}'$ and $a \in \{a_0, a_1\}$,

$$p_{s(n),s(n+1)}(a) = 1 \quad \text{for} \quad n < N - 1 \qquad \text{and} \qquad p_{s(N-1),(0,0)}(a) = 1.$$

That is, whenever the action a_0 is chosen, a transition to a pseudo state takes place. (Transition into which pseudo state depends on n.) This is followed by a sequence of "upward" transitions among the pseudo states, until $s(N - 1)$ is reached, from which a transition into state $(0, 0)$ takes place.

The pseudo states are introduced to guarantee that the state $(0, 0)$ is visited exactly once every N periods, regardless of what policy is in force. This way, a stationary policy will result in a renewal process with a renewal cycle of length N. To make this work, we also need to specify, for each state i and each action a, two cost functions, $c_0(i, a)$ and $c_1(i, a)$, and one reward function, $r(i, a)$, as follows: for $i = (n, d) \in \mathcal{S}$,

$$c_0(i, a_1) = \mathsf{E}[\Delta(n) | D(n) = d], \qquad c_1(i, a_1) = 1, \quad r(i, a_1) = \mathsf{E}[\Theta | D(n) = d];$$

and

$$c_0(i, a_0) = c_1(i, a_0) = r(i, a_0) = 0;$$

whereas, for $i \in \mathcal{S}'$ and $a \in \{a_0, a_1\}$,

$$c_0(i, a) = c_1(i, a) = r(i, a) = 0.$$

(That is, in any pseudo state, both actions incur zero cost and zero reward.)

Clearly, $-c_0(i, a)$ is the expected cost reduction associated with each inspection (see Remark 6.1), $r(i, a)$ is the expected number of defective units identified by each inspection, and $c_1(i, a)$ simply counts each inspection.

Now, given a policy π, starting from any initial state $i \in \mathcal{S} \cup \mathcal{S}'$, define the long-run averages corresponding to $c_0(i, a)$, $c_1(i, a)$, and $r(i, a)$, respectively, as follows:

$$U_\pi^k(i) = \lim_{T \to \infty} \frac{1}{T} \mathsf{E}_\pi \left[\sum_{t=0}^{T-1} c_k(i^t, a^t) | i^0 = i \right] \qquad \text{for} \qquad k = 0, 1$$

and

$$W_\pi(i) = \lim_{T \to \infty} \frac{1}{T} \mathsf{E}_\pi \left[\sum_{t=0}^{T-1} r(i^t, a^t) | i^0 = i \right].$$

Hence from renewal theory, we have, for any stationary policy π, $U_\pi^k(i)$ and $W_\pi(i)$ are independent of i, and

$$U_\pi^0(i) = V^\pi/N, \qquad U_\pi^1(i) = \mathsf{E}[I^\pi]/N, \qquad W_\pi(i) = \mathsf{E}[D^\pi]/N.$$

Therefore, the problem in (6.12) is equivalent to the following problem, with π restricted to the class of stationary policies:

$$\min_\pi \qquad U_\pi^0(i) \qquad\qquad\qquad (6.18)$$

$$\text{subject to} \qquad U_\pi^1(i) \le u/N, \quad W_\pi(i) \ge \ell/N.$$

Definition 6.8 A stationary policy is said to be a *k-randomized policy*, for a given integer $k \ge 0$, if it chooses randomized actions in at most k states. (In particular, a 0-randomized policy is just a deterministic stationary policy.)

To solve the MDP in (6.18), let us consider the following linear programming with variables $q := (q_{ia})_{i \in \mathcal{S} \cup \mathcal{S}', a \in \{a_0, a_1\}}$:

$$\min \qquad \sum_{i \in \mathcal{S}} c_0(i, a_1) q_{ia_1} \qquad\qquad\qquad (6.19)$$

$$\text{subject to} \qquad \sum_{i \in \mathcal{S}} c_1(i, a_1) q_{ia_1} \le u/N,$$

$$\sum_{i \in \mathcal{S}} r(i, a_1) q_{ia_1} \ge \ell/N,$$

$$\sum_{i \in \mathcal{S} \cup \mathcal{S}'} [p_{ij}(a_0)q_{ia_0} + p_{ij}(a_1)q_{ia_1}] = q_{ja_0} + q_{ja_1}, \quad j \in \mathcal{S} \cup \mathcal{S}',$$

$$\sum_{i \in \mathcal{S} \cup \mathcal{S}'} [q_{ia_0} + q_{ia_1}] = 1,$$

$$q_{ia} \geq 0, \quad i \in \mathcal{S} \cup \mathcal{S}', \ a \in \{a_0, a_1\}.$$

Proposition 6.9 (i) If the LP in (6.19) is infeasible, then there exists no feasible policy for the constrained MDP in (6.18) or (6.12);
(ii) let $q^* = (q_{ia}^*)$ be an optimal solution of (6.19). Then any randomized policy π, such that for $i \in \mathcal{S} \cup \mathcal{S}'$, $a \in \{a_0, a_1\}$,

$$P[\pi(i) = a] = \begin{cases} q_{ia}^*/(q_{ia_0}^* + q_{ia_1}^*) & \text{if } q_{ia_0}^* + q_{ia_1}^* > 0, \\ 1_{[a=a_1]} & \text{otherwise,} \end{cases} \tag{6.20}$$

is optimal for (6.12), where $1_{[a=a_1]}$ is an indicator function that equals 1 and 0, respectively, corresponding to $a = a_1$ and $a = a_0$.
(iii) If q^* is a *basic* optimal solution to (6.19), then (6.20) defines a 2-randomized optimal policy for (6.12).

Proof. All three parts follow directly from theorem 9.2 of Feinberg [35]. Note that here any stationary policy results in a renewal process, and hence there is only one ergodic class. Therefore, the unichain condition in [35] is satisfied. □

Intuitively, the LP in (6.19) finds the best long-run average proportion of time (q_{ia}) that should ·be spent in each state i following action a. The optimality of a 2-randomized policy follows from the properties of the LP. The solution to the LP in (6.19), however, does *not* guarantee that the 2-randomized policy will have the threshold type as characterized by a sequence of thresholds in §6.2. (In other words, the two actions need not be separated by a switching curve as characterized by the thresholds. They can be positioned all over the place in the state space.) We need one more result.

Recall that $\Delta(n, d)$ is decreasing in d (Lemma 3.10). If this decreasing is *strict*, i.e.,

$$\begin{aligned} &\mathsf{E}[c_r \Theta + \phi(n+1, \Theta) - \phi(n, \Theta) | D(n) = d] \\ > \ &\mathsf{E}[c_r \Theta + \phi(n+1, \Theta) - \phi(n, \Theta) | D(n) = d'] \end{aligned} \tag{6.21}$$

for $d' > d$, then we have the following result.

Proposition 6.10 If the decreasing of $\Delta(n, d)$ in d (in Lemma 3.10) is *strict*, i.e., (6.21) holds, then any stationary optimal policy for the MDP (6.18) (or (6.12)) is of threshold type.

Proof. By contradiction. Suppose π is an optimal policy that is *not* of threshold type. We can then follow the same approach as in the proof of

Theorem 6.7 to construct a new policy π' based on π. In particular, suppose (6.14) holds for $d' > d$. Then, $\Delta(n, d) > \Delta(n, d')$ implies that inequality (6.17) now becomes

$$\mathsf{E}[\Delta'(n)] = \mathsf{E}[\Delta(n)] - (1 - p')P'[\Delta(n, d) - \Delta(n, d')] < \mathsf{E}[\Delta(n)],$$

which implies $V^{\pi'} < V^{\pi}$. Therefore π' strictly outperforms π, contradicting the optimality of π. $\square$

It is not difficult to verify (e.g., following the proof of Lemma 3.10) that if (3.5) holds as a strict inequality, then $\Delta(n, d)$ is strictly decreasing in d, and hence (6.21) holds. Putting together all this, we have the following.

Theorem 6.11 (i) The MDP in (6.12) (or 6.18) is feasible if and only if the LP in (6.19) is feasible.
(ii) Suppose the condition in (3.5) holds as a strict inequality. Then, when (6.12) is feasible, there exists an optimal policy that is of threshold type and chooses randomized actions at no more than two states. Furthermore, this optimal policy can be obtained through solving the LP in (6.19).

Proof. Part (i) follows from the first two parts of Proposition 6.9. Part (ii) follows from the third part of Proposition 6.9 and Proposition 6.10, because the former yields an optimal 2-randomized policy and the latter guarantees that this policy must be of threshold type. $\square$

We will illustrate our results through a numerical example, considering cases both with and without constraints.

Example 6.12 Consider the individual warranty cost model, with $\mathsf{E}[C(X)] = 1$ and $\mathsf{E}[C(Y)] = 4$. Suppose $N = 9$, $c_i = 0.5$, $c_r = 1.0$, and Θ is uniformly distributed on $(0.05, 0.95)$.

Case 1: No constraint: let $u/N = 1$ and $\ell/N = 0$. The optimal policy is to continue inspection at (n, d) if and only if $d \geq d_n^*$, with $d_0^* = d_1^* = d_2^* = d_3^* = 0$, $d_4^* = d_5^* = d_6^* = 1$, and $d_7^* = d_8^* = 2$. (Hence, for example, $d_3^* = 0$ and $d_4^* = 1$ means that a minimum of 4 units needs to be inspected; if there is no defective unit found, then inspection should be terminated.) Under this policy, we have

$$\begin{aligned} \text{expected total cost} &= 17.923, \\ \text{expected number of units inspected}/N &= 0.888, \\ \text{expected number of defective units identified}/N &= 0.476. \end{aligned}$$

Case 2: The constrained case: let $u/N = 0.85$ and $\ell/N = 0.15$. The optimal policy in this case is derived from the LP specified earlier; and the strict inequality condition in (6.21) is easily verified. The threshold values here are slightly different from those in Case 1: $d_0^* = d_1^* = d_2^* = d_3^* = 0$, $d_4^* = d_5^* = 1$, and $d_6^* = d_7^* = d_8^* = 2$. That is, to continue inspection at (n, d) if and only if $d \geq d_n^*$, except at the single randomized point, $(3, 0)$, where it is optimal to continue inspection and to stop, respectively, with probabilities 0.559 and 0.441. Under this policy, we have

$$
\begin{aligned}
\text{expected total cost} &= 17.953, \\
\text{expected number of units inspected}/N &= 0.85, \\
\text{expected number of defective units identified}/N &= 0.465.
\end{aligned}
$$

Note that here the upper-bound constraint is tight, corresponding to the (single) randomized point.

6.4 An Application in Semiconductor Manufacturing

The model just presented is motivated by the production-inspection process in semiconductor wafer fabrication. The inspection facility corresponds to the so-called 'wafer probe', a stage that inspects the chips on each wafer. The upstream stage consists of the facilities that produce wafers. Each completed wafer, which carries dozens or even hundreds of chips, is sent to the probe stage, where the chips are inspected, and defective chips are identified. After that, the wafer goes into a downstream stage, where it is cut to separate the chips. The defective chips are then discarded or sent to rework, and the good ones undergo final testing/screening (e.g., speed sorting) and packaging. An important issue is to design a control procedure for the wafer probe, which explicitly accounts for capacity coordination among the stages.

In principle we can still use the sequential inspection policy of Chapter 3 for the wafer probe: each wafer is a batch of, say N, chips, with each chip being defective with probability Θ. The new dimension here is that the capacity of the probe stage usually constrains the capacity of the entire fab, because the equipment involved in the wafer probe is very expensive. Hence it is important to devise an inspection procedure at the probe stage that takes into account the capacity of its upstream and downstream stages to avoid the bottleneck situation.

Suppose the upstream stage produces wafers at a rate of λ, the probe stage inspects chips at a rate of μ, and the downstream stage processes chips at a rate of ν. The time unit is chosen to be common for all three stages.

Let π denote the sequential inspection policy at the probe stage. Wherever possible we use the same notation as in the last section. To ensure overall system stability, two conditions must be met:

(i) the maximum rate for inspecting *wafers*, $\mu/\mathsf{E}[I^\pi]$, must exceed the wafer input rate, λ. Hence,

$$\mathsf{E}[I^\pi] \leq \mu/\lambda - \epsilon := u, \qquad (6.22)$$

where $\epsilon > 0$ is a small predetermined constant and u is the upper limit in (6.1);

(ii) on the other hand, the probe stage must not let too many defective chips go by without inspection, because this may overburden the downstream stage, where identified defective chips are discarded or sent away for rework without further processing. Hence, the rate of chips sent downstream that need further work, $\lambda(N - \mathsf{E}[D^\pi])$, must not exceed what the downstream can handle, at rate ν. Thus,

$$\mathsf{E}[D^\pi] \geq N - \nu/\lambda + \epsilon := \ell, \qquad (6.23)$$

where ℓ is the lower limit in (6.1).

Remark 6.13 Unlike I^π, which depends solely on the inspection policy, D^π depends also on the intrinsic quality of the batch (i.e., wafer). Hence, in principle, the lower-bound constraint in (6.23) could become infeasible. When the lower-bound constraint does become infeasible, it means that even when the upstream stage has screened out all the defective units, the downstream stage still does not have enough capacity to process the remaining chips. Hence it is a bottleneck, and the remedy should be to either decrease the processing rate at the upstream stage (i.e., decrease the value of λ) or increase the capacity of the downstream stage (i.e., increase the value of ν).

Regarding the cost function at the probe stage, we can easily reinterpret the costs in the warranty model of Chapter 3: c_i is naturally the cost to inspect each chip; c_r can be set to zero, because the probe handles no rework. The generality of the cost function $C(\cdot)$ allows much flexibility in modeling. For example, in the simplest case, consider the additive version of C. Let $C(X)$ be the net cost (cost minus revenue) at the downstream for a good chip, and $C(Y)$ the cost for a bad chip, where X and Y can be random measures corresponding to good and bad chips; or they can be simply set as $X \equiv 1$ and $Y \equiv 0$.

Under this formulation, we have a constrained MDP model, with the constraints in (6.22) and (6.23). From the last section, we know that by solving the LP in (6.19) we can identify an optimal sequential inspection policy, which is characterized by a sequence of thresholds $\{d_n\}$, with at most two randomized thresholds.

6.5 Notes

The central problem studied in this chapter is a *constrained* MDP, with two constraints reflecting the capacity limits. We have shown that the optimality of the sequential policy of Chapter 3 still holds, with appropriate *randomization* at the thresholds.

To completely characterize the optimal policy, we need to (a) derive the thresholds and (b) determine the randomization (if any) at each threshold. In principle, this can be done by linear program (LP), following standard techniques in MDP (e.g., Derman [31], Kallenberg [47], Puterman [71], and Ross [74]). Furthermore, from known results in MDP (e.g., Beutler and Ross [7], and Feinberg [35]), we know that our constrained MDP, with two constraints, will result in no more than two randomized points (i.e., states in which actions are randomized). Indeed, based on [35], we can formulate and solve a particular LP to obtain such an optimal policy (i.e., with no more than two randomized points). The problem, however, is that in general there is no guarantee that such a policy will have the threshold structure mentioned earlier, without which it would be difficult or impractical to implement the policy. (For example, it would require an extensive "table lookup" to decide which action to take in each state).

As we have demonstrated (also see [106]), exploiting the problem structure, in particular, the K-submodularity, we can guarantee that (under a mild condition) the LP solution results in an optimal policy that is characterized by a sequence of thresholds and with randomization at no more than two thresholds.

The application in §6.4 is motivated by the operations in a wafer fab. In two previous studies, [57] and [65], a partial test approach was recommended. That is, test a sample of units in a batch. If the quality of the sample is good, test all the remaining units; otherwise, mark all the untested units as defective. (In [57], a batch is a wafer, which consists of chips as units; in [65], a batch is a set of wafers, with each wafer being a unit.) The advantage of this approach is its simplicity in satisfying the capacity constraints: because the probe stage inspects either the sample or the whole batch, a suitable sample size will guarantee that the capacity constraints are satisfied. On the other hand, this being a *single* threshold policy is in general suboptimal.

In [57], the spatial dependence of defective chips on a wafer is modeled as a Markov random field; in [65], an empirical Bayes model is used to capture lot-to-lot variability, as well as the dependence of defective units in a batch of wafers. These issues are not addressed in our model here. It appears, however, that if we restrict ourselves to a policy that either inspects all the chips on a wafer or does not inspect any (which is also the policy in [65]; i.e., each wafer is treated as a basic unit in the batch), then we can use the approach developed here to generate a sequential inspection policy. And the dependence (of defective chips) among wafers is captured by the random variable Θ (although not necessarily in the same way as in [65]). For more studies on the various aspects of a wafer fab, refer to Chen et al [16], Connors and Yao [26], Connors, Feigin, and Yao [27], and Kumar [51].

7
Coordination of Interstage Inspections

Here we continue our discussion on the interstage coordination, the subject of the last two chapters, but focus on the inspection processes at two stages in tandem. The first stage produces a batch of semiproducts that, after inspection, is sent to the second, final stage, which turns the batch into end-products, and after inspection supplies the batch to customer demand.

In terms of interfacing with customer demand (at the second stage), we adopt the same setting as in Chapter 3; in particular, the finished products will be supplied under warranties or some type of service contracts. The added dimension here is that the two stages are closely coupled. Inspection decisions at the first stage will have to take into account the inspection policy of the second stage when the batch is sent downstream. Furthermore, at the second stage, the batch carries different types of defects: those due to the first stage or the second stage or both, and the overall quality of the batch also depends on the inspection policy of the first stage. For example, if the first stage inspects more units, then there will be fewer units carrying a stage-1 defect in the batch.

As in Chapter 3, we shall focus on the optimality of a class of policies that possess a sequential structure, characterized by a sequence of thresholds. Here, again, the so-called K-submodularity plays a key role in the optimality of the threshold policies.

In §7.1 we present the two-stage model focusing on the case of an additive penalty cost function (which corresponds to individual warranties—those that applied to individual units instead of a whole batch of units). The optimal policies for both stages are developed in the next two sections: §7.2 for the first stage and §7.3 for the second stage. In particular, we show that the optimal policy at the first stage, under key Condition 7.2, has the same form as in the single-stage model of Chapter 3: it is a threshold policy characterized by a sequence of thresholds. For the second stage, the optimal policy is to give priority to those units that have not been inspected by the first stage. Furthermore, it is also characterized by a two-dimensional threshold structure. We consider the special case of constant defective rates at both stages in §7.4 and revisit the optimal policy at stage 1 in §7.5. In §7.6, we extend the results to more general cost functions.

7.1   Two-Stage Quality Control

As mentioned earlier, we consider a system of two stages in tandem. The first stage feeds batches of semiproducts into the second stage, which supplies customer demand, under some type of warranty or service contract. Specifically, a batch of N units is first processed and inspected at stage 1 and then passed on to stage 2 for processing and inspection. The defective rate at the two stages are modeled as random variables, Θ_1 and Θ_2. We focus on the inspection processes at the two stages, ignoring the production aspects, but taking into account the possibility of creating defective units. There are three types of defective units in the system: those that carry either a stage-1 defect or a stage-2 defect and those that carry both. We shall refer to the last type as '1-2 defect'. In particular, the batch inspected at stage 1 only carries stage 1 defects, and the batch inspected at stage 2 can carry all three types of defects: the defective units unidentified at stage 1 can further carry a stage-2 defect. Let Y_1, Y_2, and Y_{12} denote the lifetimes of a unit that carries one of the three types of defects; let X denote the lifetime of a good (i.e., nondefective) unit. Assume, for $i = 1, 2$, $X \geq_{st} Y_i \geq_{st} Y_{12}$. (Refer to Definition 2.1.)

Assume at both stages that inspection and repair are perfect and each identified defective unit is repaired. Each unit inspected at both stages has a lifetime X. On the other hand, if a unit has not been inspected at either stage, then its lifetime is Y_1, Y_2, Y_{12}, or X, respectively with probabilities $\Theta_1(1 - \Theta_2)$, $\Theta_2(1 - \Theta_1)$, $\Theta_1\Theta_2$, and $(1 - \Theta_1)(1 - \Theta_2)$. Hence, to model the stage-1 defect, for example (and likewise for the stage-2 defect), we can first draw a sample from the distribution of Θ_1. Given this sample value, each unit is then either defective or not with a probability equal to this sample value. This way, the dependence of all N units in the same batch is quite naturally captured through the random defective rate.

Suppose that the inspection costs are, respectively, c_{i1} and c_{i2} per unit at the two stages and the repair costs are c_{r1}, c_{r2}, and c_{r12} per unit, respectively, for the three types of defects. Assume $c_{r12} \geq c_{r1} + c_{r2}$.

Suppose the warranty cost associated with each unit is $C(Y_1)$, $C(Y_2)$, $C(Y_{12})$, or $C(X)$, depending on whether the unit is defective, and if it is, what type of defect it carries. (More general cost functions are also possible; refer to §7.6.) Naturally, assume $C(\cdot)$ is a decreasing function (of the unit's lifetime). We also assume the following relations:

$$\mathsf{E}C(Y_1) \geq \mathsf{E}C(X) + c_{r1}, \quad \mathsf{E}C(Y_2) \geq \mathsf{E}C(X) + c_{r2}, \qquad (7.1)$$

and

$$\mathsf{E}C(Y_{12}) \geq \max\{\mathsf{E}C(Y_1) + c_{r12} - c_{r1}, \ \mathsf{E}C(Y_2) + c_{r12} - c_{r2}\}. \qquad (7.2)$$

This simply give adequate incentive for repairing defective units. Note in particular that the inequalities in (7.1) and (7.2) imply

$$\mathsf{E}C(Y_{12}) \geq \mathsf{E}C(X) + c_{r12}. \qquad (7.3)$$

Furthermore, we assume that the following relation holds:

$$\mathsf{E}C(Y_{12}) + \mathsf{E}C(X) - \mathsf{E}C(Y_1) - \mathsf{E}C(Y_2) \geq 0. \qquad (7.4)$$

Note that this can be rewritten as:

$$\mathsf{E}C(Y_{12}) - \mathsf{E}C(X) \geq [\mathsf{E}C(Y_1) - \mathsf{E}C(X)] + [\mathsf{E}C(Y_2) - \mathsf{E}C(X)].$$

That is, the cost penalty (i.e., the increment above the usual service cost of a good unit) for a type-12 defect is at least as much as the cost penalty of a type 1 and a type 2 defect combined.

Our problem here is to find inspection policies for stage 1 and stage 2 to minimize the expected total cost (inspection, repair, and warranty costs) for each batch of N units at either stage. We will start with some preliminary analysis.

We are particularly interested in the optimality of certain *sequential* inspection policies at both stages. Specifically, as in Chapter 3, a sequential policy works as follows: each time a unit is inspected, the posterior (conditional) distribution of the defective rate is updated, and a decision whether to continue inspection is made accordingly. Because this applies to both stages, we will omit the indices (subscripts) for the two stages. Let $D(n)$ be the number of defectives identified from inspecting n items. Then, $[\Theta|D(n) = d]$ for some integer d: $0 \leq d \leq n$ denotes the conditional distribution of Θ, updated after n units have been inspected. From Lemma 3.9, we know this conditional distribution is monotone with respect to n and d; i.e., we have the following.

Lemma 7.1 For all n and $d \leq n$, we have

$$
\begin{aligned}
[\Theta|D(n) = d + 1] \quad &\geq_{\text{lr}} \quad [\Theta|D(n+1) = d + 1] \\
&\geq_{\text{lr}} \quad [\Theta|D(n) = d] \\
&\geq_{\text{lr}} \quad [\Theta|D(n+1) = d].
\end{aligned}
$$

In particular, $[\Theta|D(n) = d]$ is increasing in d and decreasing in n, both in the sense of the likelihood ratio ordering (and hence, also in the sense of stochastic ordering: refer to Lemma 2.3).

Now consider the inspection of a batch of N units at stage 1, the first stage. Unlike the single-stage model in Chapter 3, here the decisions at stage 1 must take into account the costs that will be incurred downstream at stage 2, and they are more complicated. In fact, the objective function that stage 1 is to optimize will include the cost to be incurred at stage 2. So in order to formulate the problem at stage 1, we first need to derive expressions for the cost to be incurred at stage 2. Let $\phi(n_1, \theta_1)$ denote the minimum expected cost to be incurred at stage 2 (including inspection, repair, and warranty costs), given that n_1 units have been inspected at stage 1 and $\Theta_1 = \theta_1$. Note that here the ϕ function, by definition, corresponds to the *optimal* policy at stage 2.

We will list two properties of the ϕ function that will result in an optimal threshold policy at stage 1. (Note, however, that these two properties only constitute a *sufficient* condition for the optimality of a threshold policy; in general, they need not be satisfied. See more discussions in §7.5, where we show that these properties are satisfied when a stage-2 defect is treated as a constant.)

Condition 7.2 (i) The function $\phi(n_1, \theta_1)$ is convex in n_1. That is, for any given θ_1 and any integer $n_1 \geq 0$,

$$\phi(n_1, \theta_1) - \phi(n_1 + 1, \theta_1) \geq \phi(n_1 + 1, \theta_1) - \phi(n_1 + 2, \theta_1).$$

(ii) The function $\phi(n_1, \theta_1)$ is K-submodular in (n_1, θ_1) with $K = c_{r1}$. That is, for $\theta_1' \geq \theta_1$,

$$[\phi(n_1 + 1, \theta_1) + \phi(n_1, \theta_1')] - [\phi(n_1, \theta_1) + \phi(n_1 + 1, \theta_1')] \geq c_{r1}(\theta_1' - \theta_1).$$

(Recall that, the notion of K-submodularity was introduced in §3.2.)

7.2   Analysis of Stage 1

Here we identify the optimal policy at stage 1 under Condition 7.2. Specifically, we assume that given n units have been inspected at stage 1 and the defective rate $\Theta_1 = \theta_1$, the expected cost to be incurred at stage 2 (following an optimal policy there), denoted $\phi(n_1, \theta_1)$, satisfies the convexity and submodularity assumed in Condition 7.2.

With these properties, the optimal policy at stage 1 is characterized by a sequence of thresholds, similar to the optimal policy in the single-stage model of Chapter 3; see Theorem 7.5. The structure of this optimal policy is a consequence of the properties listed in Lemma 7.3 and Theorem 7.4.

Indeed, the main results here, Theorems 7.4 and 7.5 and their proofs are similar to Theorems 3.12 and 3.14.

Because we are dealing exclusively with the first stage in this section, we shall omit the subscript 1, with the exception of certain primitive data such as Θ_1, c_{i1}, c_{r1}. We say stage 1 is in state (n, d), for $0 \leq n \leq N$ and $d \leq n$, if n units have been inspected of which d are found defective. Denote $V_n(d)$ as the optimal (expected) future cost, starting from (n, d), which includes the possible inspection and rework costs in *both* stages and the warranty cost for the whole batch. Let $\Phi_n(d)$ and $\Psi_n(d)$ denote the expected future costs, starting from the state (n, d), respectively, for the two actions, to stop and to continue inspection. Let $D(n)$ denote the number of defective units found in inspecting n units (at stage 1). Then

$$\Phi_n(d) = \mathsf{E}[\phi(n, \Theta_1)|D(n) = d] \tag{7.5}$$

and

$$\Psi_n(d) = c_{i1} + c_{r1}\mathsf{E}[\Theta_1|D(n) = d] + \mathsf{E}[V_{n+1}(D(n+1))|D(n) = d], \tag{7.6}$$

whereas

$$V_n(d) = \min\{\Phi_n(d), \Psi_n(d)\}, \; n < N, \quad \text{and} \quad V_N(d) = \Phi_N(d). \tag{7.7}$$

(The last equation is due to the fact that when all N units are inspected, the only option is to stop inspection.) In each state (n, d), it is optimal to continue inspection if and only if $\Psi_n(d) \leq \Phi_n(d)$. Our problem here is to find a policy that minimizes $V_0(0)$.

Lemma 7.3 Under Condition 7.2,

$$\mathsf{E}[\Phi_{n+1}(D(n+1))|D(n) = d] - \Phi_n(d) + c_{r1}\mathsf{E}[\Theta_1|D(n) = d] \tag{7.8}$$

is decreasing in d and increasing in n.

Proof. From (7.5), decreasing in d in (7.8) means that for any $d \leq d'$, the following holds:

$$\mathsf{E}[\phi(n + 1, \Theta_1) - \phi(n, \Theta_1) + c_{r1}\Theta_1|D(n) = d]$$
$$\geq \mathsf{E}[\phi(n + 1, \Theta_1) - \phi(n, \Theta_1) + c_{r1}\Theta_1|D(n) = d']. \tag{7.9}$$

Let

$$g(\theta) := \phi(n + 1, \theta) - \phi(n, \theta) + c_{r1}\theta.$$

From Condition 7.2(ii), we know that $g(\theta)$ is decreasing in θ. On the other hand, from Lemma 7.1, we know

$$[\Theta_1|D(n) = d] \leq_{\text{st}} [\Theta_1|D(n) = d']$$

for any $d \leq d'$. Hence,

$$\mathsf{E}[g(\Theta_1)|D(n) = d] \geq_{st} \mathsf{E}[g(\Theta_1)|D(n) = d'],$$

which yields the desired relation in (7.9).

Next,

$$[\Theta_1|D(n-1) = d] \geq_{st} [\Theta_1|D(n) = d],$$

following Lemma 7.1. Hence, we have

$$\mathsf{E}[\phi(n+1,\Theta_1) - \phi(n,\Theta_1) + c_{r1}\Theta_1|D(n) = d]$$
$$\geq \quad \mathsf{E}[\phi(n+1,\Theta_1) - \phi(n,\Theta_1) + c_{r1}\Theta_1|D(n-1) = d]$$
$$\geq \quad \mathsf{E}[\phi(n,\Theta_1) - \phi(n-1,\Theta_1) + c_{r1}\Theta_1|D(n-1) = d],$$

where the first inequality follows from the decreasingness of $g(\theta)$, and the second inequality follows from Condition 7.2 (i). Thus, we have shown that (7.8) is increasing in n. $\square$

Theorem 7.4 Under Condition 7.2, for stage 1,

- if it is optimal to continue inspection in state (n,d), then it is also optimal to continue inspection in state $(n,d+1)$;

- if it is optimal to stop inspection in (n,d), then it is also optimal to stop inspection in $(n+1,d)$.

In particular, $\Psi_n(d) - \Phi_n(d)$ is decreasing in d (for each given n) and increasing in n (for each given d).

Proof. Note that it is optimal to continue inspection in any state (n,d) if and only if $\Psi_n(d) \leq \Phi_n(d)$. Hence, to prove the first statement amounts to proving the following: given n, for each $d < n$,

$$\Psi_n(d) \leq \Phi_n(d) \quad \Rightarrow \quad \Psi_n(d+1) \leq \Phi_n(d+1).$$

We will prove a stronger result: that $\Psi_n(d) - \Phi_n(d)$ is decreasing in d for each given n. We argue via induction on n.

When $n = N - 1$, we have

$$\mathsf{E}[V_{n+1}(D(n+1))|D(n) = d] = \mathsf{E}[\phi(n+1,\Theta_1)|D(n) = d].$$

Hence, from (7.5) and (7.6), we have

$$\Psi_n(d) - \Phi_n(d) = c_{i1} + \mathsf{E}[c_{r1}\Theta_1 + \phi(n+1,\Theta_1) - \phi(n,\Theta_1)|D(n) = d],$$

which is decreasing in d, following Lemma 7.3. Because

$$V_n(d) = \min\{\Phi_n(d), \Psi_n(d)\},$$

we have

$$V_n(d) - \Phi_n(d) = \min\{0, \Psi_n(d) - \Phi_n(d)\}.$$

Hence, $V_n(d) - \Phi_n(d)$ is also decreasing in d.

Next, consider $n < N - 1$. As an induction hypothesis, assume that $\Psi_{n+1}(d) - \Phi_{n+1}(d)$, and hence $V_{n+1}(d) - \Phi_{n+1}(d)$, is decreasing in d. Note that

$$
\begin{aligned}
&\Psi_n(d) - \Phi_n(d) \\
={}& c_{i1} + \mathsf{E}[c_{r1}\Theta_1 + V_{n+1}(D(n+1)) - \phi(n,\Theta_1)|D(n) = d] \\
={}& c_{i1} + \mathsf{E}[c_{r1}\Theta_1 + \phi(n+1,\Theta_1) - \phi(n,\Theta_1)|D(n) = d] \\
&+ \mathsf{E}[V_{n+1}(D(n+1)) - \phi(n+1,\Theta_1)|D(n) = d].
\end{aligned}
\tag{7.10}
$$

The first expectation following the second equation is decreasing in d, following Lemma 7.3. Hence it remains to show that the second expectation is also decreasing in d. Rewrite it as follows:

$$\mathsf{E}[V_{n+1}(d + I_n(d)) - \Phi_{n+1}(d + I_n(d))], \tag{7.11}$$

where $I_n(d)$ is a binary random variable that equals 1 with probability $\mathsf{E}[\Theta_1|D(n) = d]$. Because $I_n(d)$ is stochastically increasing in d, following Lemma 7.1, the expectation in (7.11) is decreasing in d, following the induction hypothesis that $V_{n+1}(d) - \Phi_{n+1}(d)$ is decreasing in d.

To prove the second statement in the theorem, it suffices to show that $\Psi_n(d) - \Phi_n(d)$ is increasing in n, given d. Make use of (7.10). The first expectation is increasing in n, following Lemma 7.3. To argue that the expectation in (7.11) is also increasing in n, note that $I_n(d)$ is stochastically decreasing in n, following Lemma 7.1, and that $V_{n+1}(d) - \Phi_{n+1}(d)$ is decreasing in d. $\square$

Theorem 7.5 For $n = 1, ..., N$, define

$$d_n^* := \min\{d \le n : \ \Psi_n(d) \le \Phi_n(d)\}.$$

Then, under Condition 7.2,

(i) d_n^* is increasing in n;

(ii) the optimal policy for inspection at stage 1 is to inspect the units one unit at a time and stop as soon as a state (n, d) is reached with $d < d_n^*$.

Proof. Part (i) follows directly from the fact that $\Psi_n(d) - \Phi_n(d)$ is increasing in n and decreasing in d (Theorem 7.4). For part (ii), observe that from the definition of d_n^*, $d < d_n^*$ implies $\Psi_n(d) > \Phi_n(d)$. Hence, it is optimal to stop inspection in (n, d). $\square$

7.3   Optimal Policy at Stage 2

The inspection process at stage 2 is considerably different from that at stage 1. The quality of the batch depends on the inspection at stage 1, and there are several different types of defects. Suppose that the units inspected at stage 1 are marked and identified with a set **a**. The first problem we face is how to deal with these units: should they be given higher or lower priority than the units in the complement set **ā**, or should the two sets not be distinguished by priorities?

This section is divided into two parts. In the first part, we address the issue of priority, and in the second part, we present the complete optimal policy for stage 2.

7.3.1   Priority Structure

To start, consider the decision whether to inspect a particular unit at stage 2, given the defective rates at both stages are known constants: $\Theta_1 = \theta_1$ and $\Theta_2 = \theta_2$. Let $R_1(\theta_1, \theta_2)$ and $R_0(\theta_1, \theta_2)$ denote the expected costs–only for the unit in question–corresponding to the two actions (at stage 2), to inspect and not to inspect, respectively. Then, clearly,

$$\begin{aligned}
&R_1(\theta_1, \theta_2) \\
={}& c_{i2} + c_{r1}\theta_1(1 - \theta_2) + c_{r2}(1 - \theta_1)\theta_2 + c_{r12}\theta_1\theta_2 + \mathsf{EC}(X) \\
={}& c_{i2} + c_{r1}\theta_1 + c_{r2}\theta_2 + (c_{r12} - c_{r1} - c_{r2})\theta_1\theta_2 + \mathsf{EC}(X) \quad (7.12)
\end{aligned}$$

and

$$\begin{aligned}
&R_0(\theta_1, \theta_2) \\
={}& (1 - \theta_1)(1 - \theta_2)\mathsf{EC}(X) + \theta_1(1 - \theta_2)\mathsf{EC}(Y_1) \\
&+ \theta_2(1 - \theta_1)\mathsf{EC}(Y_2) + \theta_1\theta_2\mathsf{EC}(Y_{12}). \quad (7.13)
\end{aligned}$$

From these two expressions, we have

$$\begin{aligned}
&R_1(\theta_1, \theta_2) - R_0(\theta_1, \theta_2) \\
={}& c_{i2} + [\mathsf{EC}(X) - \mathsf{EC}(Y_1) + c_{r1}]\theta_1(1 - \theta_2) \\
&+ [\mathsf{EC}(X) - \mathsf{EC}(Y_2) + c_{r2}]\theta_2 \\
&+ [\mathsf{EC}(Y_2) - \mathsf{EC}(Y_{12}) + c_{r12} - c_{r2}]\theta_1\theta_2 \\
={}& c_{i2} + [\mathsf{EC}(X) - \mathsf{EC}(Y_1) + c_{r1}]\theta_1 \\
&+ [\mathsf{EC}(X) - \mathsf{EC}(Y_2) + c_{r2}](1 - \theta_1)\theta_2 \\
&+ [\mathsf{EC}(Y_1) - \mathsf{EC}(Y_{12}) + c_{r12} - c_{r1}]\theta_1\theta_2.
\end{aligned}$$

We observe that the expressions are decreasing in θ_1 and in θ_2, because the relevant coefficients of θ_1 and θ_2 are all nonpositive, taking into account

(7.1) and (7.2). Furthermore, from (7.13), we have

$$R_0(\theta_1, \theta_2) - R_0(0, \theta_2)$$
$$= \theta_1[EC(Y_1) - EC(X)]$$
$$+ \theta_1\theta_2[EC(Y_{12}) + EC(X) - EC(Y_1) - EC(Y_2)].$$

Hence, the preceding is increasing in θ_2, following (7.1) and (7.4). To summarize, we have the following.

Lemma 7.6 (i) $R_1(\theta_1, \theta_2) - R_0(\theta_1, \theta_2)$ is decreasing in θ_1 and θ_2.

(ii) $R_0(\theta_1, \theta_2) - R_0(0, \theta_2)$ is increasing in θ_2.

The following theorem assures that we only need to consider policies (at stage 2) that give priority to units in the set $\bar{\mathbf{a}}$.

Theorem 7.7 For stage 2, there exists an optimal policy that does not inspect any unit already inspected at stage 1 (i.e., those in set $\mathbf{a}$) until all the other units (in $\bar{\mathbf{a}}$) are inspected.

To better understand Theorem 7.7 and to prepare for its proof, we pursue the statement of the theorem a little further to reduce it to an inequality [in (7.14)]. Suppose $|\mathbf{a}| = n_1$ units in the batch have already been inspected at stage 1. In addition, stage 2 has inspected n_2 units from $\mathbf{a}$ and $\bar{n}_2$ units from $\bar{\mathbf{a}}$. It is important to keep in mind that the units inspected at stage 2 from $\mathbf{a}$ can only carry a stage-2 defect, because any possible stage-1 defect has already been corrected. On the other hand, the units inspected at stage 2 from $\bar{\mathbf{a}}$ can still carry all three types of defects.

We shall use $D_1(n)$ and $D_2(n)$, the number of defective units identified through inspecting n units, to update our knowledge on Θ_1 and Θ_2. Note that D_1 includes the units that carry a stage-1 defect identified at both stage 1 and stage 2, and the units that carry a 1-2 defect (identified at stage 2). Similarly, D_2 includes the units that carry a stage-2 defect or a 1-2 defect (both identified at stage 2). Note that each unit that carries a 1-2 defect is counted in both D_1 and D_2. Thus, $D_1(n_1 + \bar{n}_2) = d_1$ means that from a total of $n_1 + \bar{n}_2$ units inspected at stages 1 and 2, d_1 units are found to carry a stage-1 defect; $D_2(n_2 + \bar{n}_2) = d_2$ means that from a total of $n_2 + \bar{n}_2$ units inspected at stage 2, d_2 units are found to carry a stage 2 defect. Here $(n_2, \bar{n}_2, d_1, d_2)$ denotes the state of interest. (Note that because $n_1 = |\mathbf{a}|$ is given and remains unchanged, we do not include it in the state description.)

Let $r(\theta_1, \theta_2)$ denote the inspection and possible repair cost when an item is inspected (at stage 2), given that the defective rates at the two stages are θ_1 and θ_2, respectively. Then

$$r(\theta_1, \theta_2) = c_{i2} + c_{r1}\theta_1(1 - \theta_2) + c_{r2}(1 - \theta_1)\theta_2 + c_{r12}\theta_1\theta_2$$
$$= R_1(\theta_1, \theta_2) - EC(X).$$

Starting from the state $(n_2, \bar{n}_2, d_1, d_2)$, let $V(n_2, \bar{n}_2, d_1, d_2)$ denote the optimal expected future cost, including inspection, repair and warranty costs. Suppose the decision in state $(n_2, \bar{n}_2, d_1, d_2)$ is to continue inspection. There are two alternatives in selecting the next unit for inspection: from $\bar{\mathbf{a}}$ or from $\mathbf{a}$, with expected future costs as follows:

$$E[r(\Theta_1, \Theta_2)|D_1(n_1 + \bar{n}_2) = d_1, D_2(n_2 + \bar{n}_2) = d_2]$$
$$+E[V(n_2, \bar{n}_2 + 1, d_1 + I_1, d_2 + I_2)]$$

or

$$E[r(0, \Theta_2)|D_2(n_2 + \bar{n}_2) = d_2] + E[V(n_2 + 1, \bar{n}_2, d_1, d_2 + I_2')],$$

where I_1 is a binary random variable equal to 1 with probability $E[\Theta_1|D_1(n_1 + \bar{n}_2) = d_1]$; similarly, I_2 and I_2' are independent samples of another binary random variable that equals 1 with probability $E[\Theta_2|D_2(n_2 + \bar{n}_2) = d_2]$. Note that both Θ_2 and I_2' are independent of the event $D_1(n_1 + \bar{n}_2) = d_1$.

Therefore, to prove Theorem 7.7, it suffices to show that the cost for inspecting a unit from $\bar{\mathbf{a}}$ is smaller, i.e.,

$$E[V(n_2 + 1, \bar{n}_2, d_1, d_2 + I_2')] - E[V(n_2, \bar{n}_2 + 1, d_1 + I_1, d_2 + I_2)]$$
$$\geq \quad E[r(\Theta_1, \Theta_2)|D_1(n_1 + \bar{n}_2) = d_1, D_2(n_2 + \bar{n}_2) = d_2]$$
$$-E[r(0, \Theta_2)|D_2(n_2 + \bar{n}_2) = d_2]. \tag{7.14}$$

We establish this inequality via induction. We need the following lemma, which is elementary and can be directly verified.

Lemma 7.8 $A, B, a,$ and b are all real values.
(i) Suppose $A \geq B \Rightarrow a \geq b$ and $B - b \geq A - a$. Then

$$\min\{A, B\} - \min\{a, b\} \geq A - a.$$

(ii) Suppose $A - a \geq \Delta$ and $B - b \geq \Delta$ for some Δ. Then

$$\min\{A, B\} - \min\{a, b\} \geq \Delta.$$

Proof of Theorem 7.7 via (7.14). To lighten notation, we rewrite (7.14) as follows:

$$E[V(1, 0, 0, I_2')] - E[V(0, 1, I_1, I_2)] \tag{7.15}$$
$$\geq \quad E[r(\Theta_1, \Theta_2)|D_1(0) = d_1, D_2(0) = d_2] - E[r(0, \Theta_2)|D_2(0) = d_2].$$

That is, we effectively treat $(n_2, \bar{n}_2, d_1, d_2)$ as the starting point in calculating the future costs.

In what follows, we prove a stronger, sample-path version of the preceding:

$$V(1, 0, 0, I_2') - V(0, 1, I_1, I_2) \tag{7.16}$$
$$\geq \quad E[r(\Theta_1, \Theta_2)|D_1(0) = d_1, D_2(1) = d_2 + I_2']$$
$$-E[r(0, \Theta_2)|D_2(1) = d_2 + I_2].$$

Taking expectations on both sides of (7.16) recovers (7.15).

We will prove (7.16) by induction. Because there are uninspected units in both $\mathbf{a}$ and $\bar{\mathbf{a}}$, we have $n_2 + \bar{n}_2 \leq N - 2$. Consider the case of $n_2 + \bar{n}_2 = N - 2$. Because there is only one uninspected unit in $\mathbf{a}$ and another in $\bar{\mathbf{a}}$, we have

$$
\begin{aligned}
&V(0, 1, I_1, I_2) \\
= \ &\min\{\mathsf{E}[R_1(0, \Theta_2)|D_2(1) = d_2 + I_2] + (N - 1)\mathsf{E}C(X), \\
&\quad \mathsf{E}[R_0(0, \Theta_2)|D_2(1) = d_2 + I_2] + (N - 1)\mathsf{E}C(X)\},
\end{aligned}
$$

and

$$
\begin{aligned}
&V(1, 0, 0, I_2') \\
= \ &\min\{\mathsf{E}[R_1(\Theta_1, \Theta_2)|D_1(0) = d_1, D_2(1) = d_2 + I_2'] + (N - 1)\mathsf{E}C(X), \\
&\quad \mathsf{E}[R_0(\Theta_1, \Theta_2)|D_1(0) = d_1, D_2(1) = d_2 + I_2'] + (N - 1)\mathsf{E}C(X)\}.
\end{aligned}
$$

For convenience, write the two expressions as

$$
V(1, 0, 0, I_2') = \min\{A, B\} \quad \text{and} \quad V(0, 1, I_1, I_2) = \min\{a, b\}.
$$

We can then invoke Lemma 7.8(i). The two required conditions, $A \geq B \Rightarrow a \geq b$ and $B - b \geq A - a$, can be verified as follows from Lemma 7.6. Hence, (7.16) holds, because its right-hand side is $A - a$, taking into account the following:

$$
\begin{aligned}
&[R_1(\theta_1, \theta_2) + (N - 1)\mathsf{E}C(X)] - [R_1(0, \theta_2) + (N - 1)\mathsf{E}C(X)] \\
= \ &r(\theta_1, \theta_2) - r(0, \theta_2).
\end{aligned}
$$

As an induction hypothesis, suppose (7.16) holds for $n_2 + \bar{n}_2 \geq n + 1$. We now show it also holds for $n_2 + \bar{n}_2 = n$. Denote as in §7.2 the future costs corresponding to continuing and stopping inspection, respectively, as Ψ and Φ. (As will become evident, all we need are some differences that involve Ψ and Φ, but not Ψ and Φ themselves.) We can now express the left-hand side of (7.16) as follows:

$$
\begin{aligned}
&V(1, 0, 0, I_2') - V(0, 1, I_1, I_2) \\
= \ &\min\{\Psi(1, 0, 0, I_2'), \Phi(1, 0, 0, I_2')\} - \min\{\Psi(0, 1, I_1, I_2), \Phi(0, 1, I_1, I_2)\} \\
:= \ &\min\{A, B\} - \min\{a, b\}.
\end{aligned}
$$

Then we have

$$
\begin{aligned}
&B - b \\
:= \ &\Phi(1, 0, 0, I_2') - \Phi(0, 1, I_1, I_2) \\
= \ &\mathsf{E}[R_0(\Theta_1, \Theta_2)|D_1(0) = d_1, D_2(1) = d_2 + I_2'] \\
&\quad -\mathsf{E}[R_0(0, \Theta_2)|D_2(1) = d_2 + I_2] \\
\geq \ &\mathsf{E}[R_1(\Theta_1, \Theta_2)|D_1(0) = d_1, D_2(1) = d_2 + I_2']
\end{aligned}
$$

$$-\mathsf{E}[R_1(0,\Theta_2)|D_2(1) = d_2 + I_2]$$
$$= \quad \mathsf{E}[r(\Theta_1,\Theta_2)|D_1(0) = d_1, D_2(1) = d_2 + I_2']$$
$$-\mathsf{E}[r(0,\Theta_2)|D_2(1) = d_2 + I_2]$$
$$:= \quad \Delta,$$

where the inequality follows from Lemma 7.6. In addition,

$$A - a$$
$$:= \quad \Psi(1,0,0,I_2') - \Psi(0,1,I_1,I_2)$$
$$= \quad \mathsf{E}[r(\Theta_1,\Theta_2)|D_1(0) = d_1, D_2(1) = d_2 + I_2']$$
$$+\mathsf{E}[V(1,1,I_1,I_2 + I_2')]$$
$$-\mathsf{E}[r(\Theta_1,\Theta_2)|D_1(1) = d_1 + I_1, D_2(1) = d_2 + I_2]$$
$$-\mathsf{E}[V(0,2,I_1,I_2 + I_2'')]$$
$$\geq \quad \mathsf{E}[r(\Theta_1,\Theta_2)|D_1(0) = d_1, D_2(1) = d_2 + I_2']$$
$$+\mathsf{E}[V(1,1,I_1,I_2 + I_2')]$$
$$-\mathsf{E}[r(0,\Theta_2)|D_2(1) = d_2 + I_2]$$
$$-\mathsf{E}[V(1,1,I_1,I_2 + I_2')]$$
$$= \quad \mathsf{E}[r(\Theta_1,\Theta_2)|D_1(0) = d_1, D_2(1) = d_2 + I_2']$$
$$-\mathsf{E}[r(0,\Theta_2)|D_2(1) = d_2 + I_2]$$
$$= \quad \Delta,$$

where the inequality follows from the induction hypothesis. (I_2'' is another independent sample from the distribution of I_2 and I_2'.) The desired claim follows from Lemma 7.8 (ii), noticing that Δ is exactly the right-hand side of (7.16). $\square$

7.3.2   Threshold Structure

From the preceding discussion, note that in general a full state description for the control at stage 2 is $(n_2, \bar{n}_2, d_1, d_2)$, where n_2 and $\bar{n}_2$ denote the number of units inspected from **a** and **ā**, and d_1 and d_2 denote the number of stage-1 and stage-2 defectives identified respectively. The immediate consequence of Theorem 7.7 is a simplification of the state description. Let n_2 now denote the *total* number of units stage 2 has inspected. Then there is no ambiguity: if $n_2 \leq N - |\mathbf{a}|$, then all these units are from **ā**; otherwise, $N - |\mathbf{a}|$ are from **ā** and the remaining are from **a**. So we remove $\bar{n}_2$ from the state description, and let (n_2, d_1, d_2) denote the new state. We will continue to denote $n_1 := |\mathbf{a}|$.

Let $\phi(n_2, \theta_1, \theta_2)$ denote the expected future cost if stage 2 stops inspection after inspecting a total of n_2 units, given $\Theta_1 = \theta_1$ and $\Theta_2 = \theta_2$. Note that $\phi(n_2, \theta_1, \theta_2)$ is the warranty cost of all the N units, n_2 of which have been inspected at stage 2, while the other $N - n_2$ units have not been inspected. Following Theorem 7.7, there are two cases: for $n_2 \leq N - n_1 - 1$,

$$\phi(n_2, \theta_1, \theta_2)$$
$$= n_2 \mathsf{E}C(X) + n_1 R_0(0, \theta_2) + (N - n_1 - n_2) R_0(\theta_1, \theta_2); \quad (7.17)$$

and for $n_2 \geq N - n_1$,

$$\phi(n_2, \theta_1, \theta_2) = n_2 \mathsf{E}C(X) + (N - n_2) R_0(0, \theta_2). \quad (7.18)$$

In the first case, all the units inspected are from $\bar{\mathbf{a}}$. The uninspected units then include the n_1 units in $\mathbf{a}$ and the remaining $N - n_1 - n_2$ units in $\bar{\mathbf{a}}$. In the second case, all the units in $\bar{\mathbf{a}}$ have been inspected. In addition, some units (possibly none) in $\mathbf{a}$ have also been inspected. Hence, the $N - n_2$ uninspected units are all in $\mathbf{a}$.

In state (n_2, d_1, d_2), let $V_{n_2}(d_1, d_2)$ be the expected total future cost. Let $\Phi_{n_2}(d_1, d_2)$ and $\Psi_{n_2}(d_1, d_2)$ be the expected total future costs corresponding to stopping or continuing inspection. We have

$$V_{n_2}(d_1, d_2) = \min\{\Phi_{n_2}(d_1, d_2), \Psi_{n_2}(d_1, d_2)\}, \; n_2 < N; \quad (7.19)$$
$$V_N(d_1, d_2) = \Phi_N(d_1, d_2)$$

where

$$\Phi_{n_2}(d_1, d_2) = \mathsf{E}[\phi(n_2, \Theta_1, \Theta_2)|D_1(n_1 + n_2) = d_1, D_2(n_2) = d_2] \quad (7.20)$$

and

$$\Psi_{n_2}(d_1, d_2) = \mathsf{E}[r(\Theta_1, \Theta_2)|D_1(n_1 + n_2) = d_1, D_2(n_2) = d_2]$$
$$+ \mathsf{E}[V_{n_2+1}(d_1 + I_1, d_2 + I_2)] \quad (7.21)$$

for $n_2 \leq N - n_1 - 1$, and

$$\Psi_{n_2}(d_1, d_2) = \mathsf{E}[r(0, \Theta_2)|D_2(n_2) = d_2] + \mathsf{E}[V_{n_2+1}(d_1, d_2 + I_2)] \quad (7.22)$$

for $n_2 \geq N - n_1$. Note, in these expressions,

$$I_1 := \mathbf{1}(\mathsf{E}[\Theta_1|D_1(n_1 + n_2) = d_1]), \quad I_2 := \mathbf{1}(\mathsf{E}[\Theta_2|D_2(n_2) = d_2]), \quad (7.23)$$

where $\mathbf{1}(p)$ denotes a binary (0-1) random variable that equals one with probability p.

The problem here is to find an inspection policy for stage 2 to minimize $V_0(D_1(n_1), 0)$ for any value of $D_1(n_1)$, given $|\mathbf{a}| = n_1$.

The following lemma is analogous to Lemma 7.3 and plays a key role in establishing the optimality of the threshold policy for stage 2.

Lemma 7.9 The following expressions are decreasing in d_1 and d_2 and increasing in n_2:

$$E[\Phi_{n_2+1}(d_1 + I_1, d_2 + I_2)] - \Phi_{n_2}(d_1, d_2)$$
$$+E[r(\Theta_1, \Theta_2)|D_1(n_1 + n_2) = d_1, D_2(n_2) = d_2] \qquad (7.24)$$

for $n_2 \le N - n_1 - 1$; and

$$E[\Phi_{n_2+1}(d_1 + I_1, d_2 + I_2)] - \Phi_{n_2}(d_1, d_2)$$
$$+E[r(0, \Theta_2)|D_1(n_1 + n_2) = d_1, D_2(n_2) = d_2] \qquad (7.25)$$

for $n_2 \ge N - n_1$; where I_1 and I_2 follow the definitions in (7.23).

Proof. The basic idea is similar to the proof of Lemma 7.3. First, for $n_2 \le N - n_1 - 1$, let

$$\begin{aligned}
G_1(\theta_1, \theta_2) \quad &:= \quad \phi(n_2 + 1, \theta_1, \theta_2) - \phi(n_2, \theta_1, \theta_2) + r(\theta_1, \theta_2) \\
&= \quad EC(X) - R_0(\theta_1, \theta_2) + r(\theta_1, \theta_2) \\
&= \quad R_1(\theta_1, \theta_2) - R_0(\theta_1, \theta_2).
\end{aligned}$$

Then $G_1(\theta_1, \theta_2)$ is decreasing in θ_1 and θ_2, following Lemma 7.6. On the other hand, from Lemma 7.1, we know that given n_1 and n_2, $[\Theta_1|D_1(n_1 + n_2) = d_1]$ is stochastically increasing in d_1, and $[\Theta_2|D_2(n_2) = d_2]$ is stochastically increasing in d_2. Hence,

$$E[G_1(\Theta_1, \Theta_2)|D_1(n_1 + n_2) = d_1, D_2(n_2) = d_2] \qquad (7.26)$$

is decreasing in d_1 and d_2, and so is the expression in (7.24).

Similarly, for $n_2 \ge N - n_1$, let

$$\begin{aligned}
G_0(\theta_1, \theta_2) \quad &:= \quad \phi(n_2 + 1, \theta_1, \theta_2) - \phi(n_2, \theta_1, \theta_2) + r(0, \theta_2) \\
&= \quad R_1(0, \theta_2) - R_0(0, \theta_2) \\
&= \quad \theta_2[EC(X) - EC(Y_2) + c_{r2}].
\end{aligned}$$

Then $G_0(\theta_1, \theta_2)$ is independent of θ_1, and decreasing in θ_2, taking into account (7.1). Repeating the earlier argument proves that

$$E[G_0(\Theta_1, \Theta_2)|D_1(n_1 + n_2) = d_1, D_2(n_2) = d_2] \qquad (7.27)$$

is also decreasing in d_1 and d_2; hence so is the expression in (7.25).

To prove that (7.24) and (7.25) are increasing in n_2, note that both G_1 and G_0 are independent of n_2 and (again following Lemma 7.1) that given d_1 and d_2, $[\Theta_1|D_1(n_1 + n_2) = d_1]$ and $[\Theta_2|D_2(n_2) = d_2]$ are stochastically decreasing in n_2. Thus, both (7.26) and (7.27) are increasing in n_2; and hence so are (7.24) and (7.25). $\square$

Note that the proof also establishes the K-submodularity of the function $\phi(n_2, \theta_1, \theta_2)$.

Corollary 7.10 (i) For $n_2 \leq N - n_1 - 1$, $\phi(n_2, \theta_1, \theta_2)$ is K_1-submodular with respect to (n_2, θ_1), with

$$K_1 = c_{r1} + (c_{r12} - c_{r1} - c_{r2})\theta_2,$$

and K_2-submodular with respect to (n_2, θ_2), with

$$K_2 = c_{r2} + (c_{r12} - c_{r1} - c_{r2})\theta_1.$$

(ii) For all $n_2 \geq N - n_1$, $\phi(n_2, \theta_1, \theta_2)$ is K_1-submodular with respect to (n_2, θ_1), with $K_1 = 0$ (i.e., submodular), and K_2-submodular with respect to (n_2, θ_2), with $K_2 = c_{r2}$.

Proof. The claimed properties follow from the increasingness in θ_1 and θ_2 of $G_1(\theta_1, \theta_2)$ and $G_0(\theta_1, \theta_2)$ in the proof of Lemma 7.9, taking into account the following relation:

$$
\begin{aligned}
r(\theta_1, \theta_2) &= K_1\theta_1 + c_{r2}\theta_2 + c_{i2} \\
&= K_2\theta_2 + c_{r1}\theta_1 + c_{i2}. \qquad \square
\end{aligned}
$$

Theorem 7.11 For stage 2,

- if it is optimal to continue inspection in a state (n_2, d_1, d_2), then it is also optimal to continue inspection in any states (n_2, d_1', d_2'), with $d_1' \geq d_1$ and $d_2' \geq d_2$;

- if it is optimal to stop inspection in a state (n_2, d_1, d_2), then it is also optimal to stop inspection in any states (n_2', d_1, d_2), with $n_2' \geq n_2$.

In particular, $\Psi_{n_2}(d_1, d_2) - \Phi_{n_2}(d_1, d_2)$ is decreasing in d_1 and d_2 and increasing in n_2.

Proof. Proceed in the same way as in the proof of Theorem 7.4. Specifically, we want to prove via induction that $\Psi_{n_2}(d_1, d_2) - \Phi_{n_2}(d_1, d_2)$ is decreasing in d_1 and d_2 and increasing in n_2. Here, in place of (7.10), we have

$$
\begin{aligned}
&\Psi_{n_2}(d_1, d_2) - \Phi_{n_2}(d_1, d_2) \\
=\ &\mathsf{E}[r(\Theta_1, \Theta_2)|D_1(n_1 + n_2) = d_1, D_2(n_2) = d_2] \\
&+\Phi_{n_2+1}(d_1 + I_1, d_2 + I_2) - \Phi_{n_2}(d_1, d_2) \\
&+\mathsf{E}[V_{n_2+1}(d_1 + I_1, d_2 + I_2) - \Phi_{n_2+1}(d_1 + I_1, d_2 + I_2)] \quad (7.28)
\end{aligned}
$$

for $n_2 \leq N - n_1 - 1$ and

$$
\begin{aligned}
&\Psi_{n_2}(d_1, d_2) - \Phi_{n_2}(d_1, d_2) \\
=\ &\mathsf{E}[r(0, \Theta_2)|D_2(n_2) = d_2] \\
&+\Phi_{n_2+1}(d_1 + I_1, d_2 + I_2) - \Phi_{n_2}(d_1, d_2) \\
&+\mathsf{E}[V_{n_2+1}(d_1 + I_1, d_2 + I_2) - \Phi_{n_2+1}(d_1 + I_1, d_2 + I_2)] \quad (7.29)
\end{aligned}
$$

for $n_2 \geq N - n_1$. The rest is the same as in the proof of Theorem 7.4. In particular, the desired monotonicity properties (in d_1, d_2, and n_2) follow from Lemma 7.9. $\square$

Given n_1, define

$$S_{n_1}(n_2) \quad := \quad \{(d_1, d_2): \ 0 \leq d_1 \leq n_1 + n_2, \ 0 \leq d_2 \leq n_2;$$
$$\Psi_{n_2}(d_1, d_2) > \Phi_{n_2}(d_1, d_2)\}. \tag{7.30}$$

The following properties of $S_{n_1}(n_2)$ are direct consequences of Theorem 7.11.

Corollary 7.12 The set $S_{n_1}(n_2)$ in (7.30) satisfies the following properties:

(i) it is a lower set, i.e., $(d_1, d_2) \in S_{n_1}(n_2) \Rightarrow (d_1', d_2') \in S_{n_1}(n_2)$, for any $d_1' \leq d_1$ and $d_2' \leq d_2$;

(ii) $S_{n_1}(n_2) \subseteq S_{n_1}(n_2')$ for any $n_2' \geq n_2$.

In view of (i), the boundary of $S_{n_1}(n_2)$ is a set of points (two-dimensional vectors), each being a nondominant point in $S_{n_1}(n_2)$, denoted (d_1^*, d_2^*). That is, for any point $(d_1, d_2) \in S_{n_1}(n_2)$, $d_1 \geq d_1^*$ and $d_2 \geq d_2^*$ implies $(d_1, d_2) \equiv (d_1^*, d_2^*)$.

To summarize, we have the following.

Theorem 7.13 The optimal inspection policy at stage 2 is as follows:

(i) first inspect the units in $\bar{\mathbf{a}}$, i.e., those that have not yet been inspected at stage 1. Inspect the units in $\mathbf{a}$ only after all units in $\bar{\mathbf{a}}$ have been inspected;

(ii) stop inspection as soon as a state (n_2, d_1, d_2) with $(d_1, d_2) \in S_{n_1}(n_2)$ is reached.

To fully characterize the optimal policy at stage 2, for each n_1 value (the number of units inspected at stage 1), first compute $\Psi_{n_2}(d_1, d_2)$ and $\Phi_{n_2}(d_1, d_2)$ recursively from the expressions presented earlier, (7.17) through (7.23). Next, derive the set $S_{n_1}(n_2)$. From Corollary 7.12, $S_{n_1}(n_2)$ of (7.30) can be derived as follows: for each $0 \leq d_2 \leq n_2$, derive

$$d_1^*(d_2) := \min\{d_1 : \Psi_{n_2}(d_1, d_2) \leq \Phi_{n_2}(d_1, d_2)\}; \tag{7.31}$$

then

$$S_{n_1}(n_2) = \{(d_1, d_2) : 0 \leq d_2 \leq n_2, 0 \leq d_1 < d_1^*(d_2)\}. \tag{7.32}$$

7.4   A Special Case: Constant Defective Rates

Here we consider the special case in which both stages have constant defective rates: $\Theta_1 \equiv \theta_1$ and $\Theta_2 \equiv \theta_2$. This special case was studied in some earlier works (e.g., Lindsay and Bishop [55]). Here, in addition to showing that the optimality of either 0% or 100% inspection at each stage follows directly from our model, we derive the threshold values that determine the switchover between inspecting all units or nothing at all.

The optimality of this "all-or-nothing" policy should follow rather intuitively from the constant defective rates. Because in this case there is no "learning" in the sense of updating the conditional defective rate as inspection progresses, if an action is optimal in one state, it should be optimal in all states. In other words, the optimal policy in each stage should be either to inspect all units or not to inspect any unit.

Consider first the optimal policy at stage 2, given that n_1 units have been inspected at stage 1. Then the total expected cost at stage 2 given n_2 units are inspected can be expressed as follows:

$$n_2 R_1(\theta_1, \theta_2) + n_1 R_0(0, \theta_2) + (N - n_1 - n_2) R_0(\theta_1, \theta_2)$$
$$= n_2[R_1(\theta_1, \theta_2) - R_0(\theta_1, \theta_2)] + n_1 R_0(0, \theta_2)$$
$$+ (N - n_1) R_0(\theta_1, \theta_2) \tag{7.33}$$

for $n_2 \leq N - n_1$ and

$$(N - n_1) R_1(\theta_1, \theta_2) + (n_1 + n_2 - N) R_1(0, \theta_2) + (N - n_2) R_0(0, , \theta_2)$$
$$= n_2[R_1(0, \theta_2) - R_0(0, \theta_2)] + (N - n_1)[R_1(\theta_1, \theta_2) - R_1(0, \theta_2)]$$
$$+ N R_0(0, , \theta_2) \tag{7.34}$$

for $N - n_1 \leq n_2 \leq N$. Hence, to minimize the costs, the solution is:

- $n_2^* = 0$, if $R_1(\theta_1, \theta_2) > R_0(\theta_1, \theta_2)$,

- $n_2^* = N$, if $R_1(0, \theta_2) \leq R_0(0, \theta_2)$, or

- $n_2^* = N - n_1$, if $R_1(\theta_1, \theta_2) \leq R_0(\theta_1, \theta_2)$ and $R_1(0, \theta_2) > R_0(0, \theta_2)$.

From (7.12) and (7.13), it is easy to verify that $R_1(0, \theta_2) \leq R_0(0, \theta_2)$ is equivalent to

$$\theta_2 \geq \frac{c_{i2}}{\mathsf{E}C(Y_2) - \mathsf{E}C(X) - c_{r2}} := \bar{\theta}_2; \tag{7.35}$$

and $R_1(\theta_1, \theta_2) > R_0(\theta_1, \theta_2)$ is equivalent to

$$\theta_2$$
$$< \frac{c_{i2} - \theta_1[\mathsf{E}C(Y_1) - \mathsf{E}C(X) - c_{r1}]}{\theta_1[\mathsf{E}C(Y_{12}) - \mathsf{E}C(Y_1) - c_{r12} + c_{r1}] + (1 - \theta_1)[\mathsf{E}C(Y_2) - \mathsf{E}C(X) - c_{r2}]}$$
$$:= \underline{\theta}_2. \tag{7.36}$$

Summing up the discussion, we have the following.

Lemma 7.14 Suppose both stages have constant defective rates: $\Theta_1 \equiv \theta_1$ and $\Theta_2 \equiv \theta_2$. Suppose stage 1 has inspected a total of $n_1 = |\mathbf{a}|$ units. Then the optimal policy at stage 2 is:

(i) to inspect all N units, if $\theta_2 \geq \bar{\theta}_2$ in (7.35);

(ii) not to inspect at all, if $\theta_2 < \underline{\theta}_2$ in (7.36); and

(iii) to inspect the $N - n_1$ units in $\bar{\mathbf{a}}$ only, if $\bar{\theta}_2 > \theta_2 \geq \underline{\theta}_2$.

Proof. The only point left out of the discussion is that in (iii) we need to justify that $\bar{\theta}_2 \geq \underline{\theta}_2$. But this can be directly verified from comparing (7.35) and (7.36), along with (7.1) and (7.2), taking into account that $\bar{\theta}_2 \leq 1$. $\square$

The optimal costs corresponding to the three cases in Lemma 7.14 are, respectively:

$$\phi_1 = n_1 R_1(0, \theta_2) + (N - n_1) R_1(\theta_1, \theta_2), \tag{7.37}$$
$$\phi_2 = n_1 R_0(0, \theta_2) + (N - n_1) R_0(\theta_1, \theta_2), \tag{7.38}$$
$$\phi_3 = n_1 R_0(0, \theta_2) + (N - n_1) R_1(\theta_1, \theta_2). \tag{7.39}$$

Hence, the total expected cost at stage 1, given n_1 units are inspected, can be expressed as

$$n_1(c_{i1} + \theta_1 c_{r1}) + \phi^*(n_1, \theta_1, \theta_2), \tag{7.40}$$

where ϕ^* is equal to ϕ_1, ϕ_2, or ϕ_3 according to which range θ_2 falls into in Lemma 7.14. Regardless, however, ϕ^* is linear in n_1. Hence, the expression in (7.40) is also a linear function of n_1; and the optimal n_1 that minimizes it is either zero or N. Hence, we only need to compare these two alternatives. Observe from the discussion that at stage 1, inspecting N units is better than inspecting no units if and only if

$$N(c_{i1} + \theta_1 c_{r1}) + \min\{N R_1(0, \theta_2), N R_0(0, \theta_2)\}$$
$$\leq \min\{N R_1(\theta_1, \theta_2), N R_0(\theta_1, \theta_2)\}.$$

This is equivalent to:

$$R_1(\theta_1, \theta_2) - \theta_1 c_{r1} \geq c_{i1} + \min\{R_1(0, \theta_2), R_0(0, \theta_2)\}, \tag{7.41}$$
$$R_0(\theta_1, \theta_2) - \theta_1 c_{r1} \geq c_{i1} + \min\{R_1(0, \theta_2), R_0(0, \theta_2)\}. \tag{7.42}$$

Now subtract $R_1(0, \theta_2)$ from both sides of (7.41). The left-hand side becomes

$$R_1(\theta_1, \theta_2) - R_1(0, \theta_2) - \theta_1 c_{r1} = [R_1(1, \theta_2) - R_1(0, \theta_2) - c_{r1}]\theta_1,$$

taking into account (7.12). Also note that the quantity in the squared brackets is nonnegative. Hence, from (7.41), we obtain

$$\theta_1 \geq \frac{c_{i1} + \min\{R_0(0, \theta_2) - R_1(0, \theta_2), 0\}}{R_1(1, \theta_2) - R_1(0, \theta_2) - c_{r1}} := \theta'. \tag{7.43}$$

Note that when the denominator is equal to zero, θ' is understood to be 1. Similarly, from (7.42), we obtain

$$\theta_1 \geq \frac{c_{i1} + \min\{R_1(0, \theta_2) - R_0(0, \theta_2), 0\}}{R_0(1, \theta_2) - R_0(0, \theta_2) - c_{r1}} := \theta'', \qquad (7.44)$$

with the understanding that $\theta'' = 1$ should the denominator become zero.

Therefore, the threshold value at stage 1 should be set at $\theta_1^* = \max\{\theta', \theta''\}$: inspect all N units if $\theta_1 \geq \theta_1^*$; otherwise, inspect nothing. Note that if the denominator in either (7.43) or (7.44) is equal to zero, then inspect N units only if $\theta_1 \equiv 1$.

Once the optimal decision at stage 1 is carried out, it impacts the decision at stage 2 following Lemma 7.14. Specifically, part (iii) of the lemma should be combined into the first two parts as follows: if stage 1 inspects all N units, i.e., $n_1 = |\mathbf{a}| = N$, then it is optimal to inspect all N units if $\theta_2 \geq \bar{\theta}_2$ and to inspect 0 units if $\theta_2 < \bar{\theta}_2$. If $n_1 = |\mathbf{a}| = 0$, then it is optimal to inspect all N units if $\theta_2 \geq \underline{\theta}_2$ and inspect 0 units if $\theta_2 < \underline{\theta}_2$.

To summarize, we have the following.

Theorem 7.15 Suppose both stages have constant defective rates: $\Theta_1 \equiv \theta_1$ and $\Theta_2 \equiv \theta_2$.

(i) At stage 1, it is optimal to inspect all N units if $\theta_1 \geq \theta_1^*$ and to inspect 0 units if $\theta_1 < \theta_1^*$, where $\theta_1^* = \max\{\theta', \theta''\}$, with θ', θ'' following (7.43 and 7.44).

(ii) At stage 2, it is optimal to inspect all N units if $\theta_2 \geq \theta_2^*$ and to inspect 0 units if $\theta_2 < \theta_2^*$, where $\theta_2^* = \bar{\theta}_2$ or $\underline{\theta}_2$, according to, respectively, $\theta_1 \geq \theta_1^*$ or $\theta_1 < \theta_1^*$, with $\bar{\theta}_2$ and $\underline{\theta}_2$ following (7.35 and 7.36).

7.5   Optimal Policy at Stage 1

We now revisit the characterization of the optimal policy at stage 1. Recall in §7.1 that we established, under Condition 7.2, that there exists an optimal policy at stage 1 with the threshold structure described in Theorem 7.5. Here, let us examine more closely the derivation of the optimal policy at stage 1 via dynamic programming as outlined in §7.2. The starting point is to compute the function in (7.5), which we rewrite as follows, recovering the stage indices,

$$\Phi_{n_1}(d_1) = \mathsf{E}[\phi(n_1, \Theta_1) | D_1(n_1) = d_1],$$

which, in turn, is equal to the optimal value at stage 2: $V_0(D_1(n_1), 0)$, with $D_1(n_1) = d_1$; refer to the dynamic programming formulation of the stage-2 problem in §7.3.2 (preceding Lemma 7.9). In other words, for each (d_1, n_1), with $d_1 \leq n_1 \leq N$, we first solve the stage-2 problem, which returns the functional value in (7.5). Then we go through the recursion in (7.5) through (7.7) to generate the optimal actions in stage 1 for each state (d_1, n_1).

This procedure applies, with or without Condition 7.2. In other words, even without Condition 7.2, the dynamic programming problem for stage 1 has already significantly benefited from the structure of the optimal policy at stage 2. With Condition 7.2, there is an additional advantage in the dynamic programming recursion: for each n, as we increase the d value and compare $\Psi_n(d)$ with $\Phi_n(d)$, we can stop as soon as $\Psi_n(d) \leq \Phi_n(d)$ (which will continue to hold for larger d values, as guaranteed by the threshold structure). This reduces the computational effort by roughly half.

Also note that Condition 7.2 is only a sufficient condition: without it, the optimal actions may still follow a threshold structure. In general, the convexity and submodularity properties required in Condition 7.2 need not be satisfied. We will show, however, that these properties are indeed satisfied in a special case: when the defective rate at stage 2 is a constant.

Theorem 7.16 Suppose stage 2 has a constant defective rate: $\Theta_2 \equiv \theta_2$. Then Condition 7.2 is satisfied. Specifically, the function $\phi(n_1, \theta_1)$ (optimal cost at stage 2, given n_1 units are inspected at stage 1 and $\Theta_1 = \theta_1$) is linear in n_1, and K-submodular in (n_1, θ_1) with $K = c_{r1}$. Consequently, the optimal policy at stage 1 has the threshold structure described in Theorem 7.5.

Proof. Under the stated conditions, $\phi(n_1, \theta_1)$ is just the ϕ^* function following Lemma 7.14. Hence the linearity in n_1 follows from the discussion there.

For K-submodularity, we want to show that for $\theta_1' \geq \theta_1$,

$$
\begin{aligned}
\text{LHS} \quad &:= \quad [\phi(n_1 + 1, \theta_1) + \phi(n_1, \theta_1')] - [\phi(n_1, \theta_1) + \phi(n_1 + 1, \theta_1')] \\
&\geq \quad c_{r1}(\theta_1' - \theta_1).
\end{aligned}
$$

Consider the three cases in Lemma 7.14.

Case (i). The optimal solution is ϕ_1 at θ_1, and it is also optimal at θ_1'. Hence, following (7.37), we have

$$
\begin{aligned}
\text{LHS} \quad &= \quad R_1(\theta_1', \theta_2) - R_1(\theta_1, \theta_2) \\
&= \quad c_{r1}(\theta_1' - \theta_1) + (c_{r12} - c_{r1} - c_{r2})(\theta_1' - \theta_1)\theta_2 \\
&\geq \quad c_{r1}(\theta_1' - \theta_1),
\end{aligned}
$$

where the inequality follows from the fact that $c_{r12} - c_{r1} - c_{r2} \geq 0$.

Case (ii). At θ_1, the optimal solution is ϕ_2; whereas at θ_1', the optimal solution could be either ϕ_2 or ϕ_3, because increasing θ_1 decreases the threshold value in (7.36).

First suppose at θ_1' that ϕ_2 is optimal. Then, from (7.38), we have

LHS
$$\begin{aligned}
&= R_0(\theta_1', \theta_2) - R_0(\theta_1, \theta_2) \\
&= (\theta_1' - \theta_1)[EC(Y_1) - EC(X)] \\
&\quad + \theta_2(\theta_1' - \theta_1)[EC(Y_{12}) + EC(X) - EC(Y_1) - EC(Y_2)] \\
&\geq (\theta_1' - \theta_1)[EC(Y_1) - EC(X)] \\
&\geq (\theta_1' - \theta_1)c_{r1},
\end{aligned}$$

where the last inequality follows from (7.1) and (7.4).

Next, suppose at θ_1', ϕ_3 is optimal. Then, from (7.38) and (7.39), we have

$$\begin{aligned}
\text{LHS} &= R_1(\theta_1', \theta_2) - R_0(\theta_1, \theta_2) \\
&\geq R_1(\theta_1', \theta_2) - R_1(\theta_1, \theta_2) \\
&= c_{r1}(\theta_1' - \theta_1) + (c_{r12} - c_{r1} - c_{r2})(\theta_1' - \theta_1)\theta_2 \\
&\geq c_{r1}(\theta_1' - \theta_1),
\end{aligned}$$

where the first inequality follows from $\phi_2 \leq \phi_3$, because ϕ_2 is optimal at θ_1 [cf. (7.38, and 7.39)].

Case (iii). The optimal solution is ϕ_3 at θ_1. It is also optimal at θ_1', because increasing θ_1 only decreases the lower threshold value—the right-hand side in (7.36). Hence, following (7.39), we have, as in Case (i),

$$\begin{aligned}
\text{LHS} &= R_1(\theta_1', \theta_2) - R_1(\theta_1, \theta_2) \\
&= c_{r1}(\theta_1' - \theta_1) + (c_{r12} - c_{r1} - c_{r2})(\theta_1' - \theta_1)\theta_2 \\
&\geq c_{r1}(\theta_1' - \theta_1). \quad \square
\end{aligned}$$

Note that when $\Theta_2 \equiv \theta_2$, $d_1^*(d_2)$ in (7.31) is independent of d_2, because both $\Psi_{n_2}(d_1, d_2)$ and $\Phi_{n_2}(d_1, d_2)$ are independent of d_2. Hence, we have the following.

Corollary 7.17 Suppose stage 2 has a constant defective rate. Then

(i) for stage 1 the policy described in Theorem 7.5 is optimal;

(ii) for stage 2, given $|\mathbf{a}| = n_1$, there exists a d_{n_1, n_2} for each $n_2 \leq N$, such that it is optimal to stop inspection at (n_2, d_1, d_2) if and only if $d_1 < d_{n_1, n_2}$.

Based on Corollary 7.17, we can develop a heuristic policy as follows: first, assume that stage 2 has a constant defective rate, $\Theta_2 \equiv E[\Theta_2]$, and derive a threshold policy for stage 1 as in Theorem 7.5; next, for stage 2, continue to treat the defect rate as what it is, i.e., a random variable, and derive the optimal threshold policy following Theorem 7.13.

n_1	0	1	2	3	4	5	6	7	8	9	10	11	12	13	14
d_{n_1}	0	1	1	2	2	3	4	4	5	5	6	7	7	8	8
n_1	15	16	17	18	19	20	21	22	23	24	25	26	27	28	29
d_{n_1}	9	10	10	11	11	12	13	13	13	14	14	15	15	15	14

TABLE 7.1. Threshold values (optimal) at stage 1.

n_1	0	1	2	3	4	5	6	7	8	9	10	11	12	13	14
d_{n_1}	0	0	0	1	1	2	2	3	3	4	4	4	5	5	6
n_1	15	16	17	18	19	20	21	22	23	24	25	26	27	28	29
d_{n_1}	6	7	7	8	8	9	9	10	10	11	11	12	12	13	14

TABLE 7.2. Threshold values (heuristics) at stage 1.

We will study this heuristic policy in two examples and compare its performance against the optimal policy. In addition, we compare it against the solution obtained by treating the defective rates at both stages as constant, i.e., assuming $\Theta_1 \equiv E[\Theta_1]$ and $\Theta_2 \equiv E[\Theta_2]$.

Example 7.18 Suppose $N = 30$; $c_{i1} = 6.5$, $c_{i2} = 13.5$; $c_{r1} = 1.0$, $c_{r2} = 2.0$, $c_{r12} = 6.0$; $EC(X) = 3.5$, $EC(Y_1) = 18.5$, $EC(Y_2) = 20$, $EC(Y_{12}) = 35.5$; and Θ_1 and Θ_2 are both uniformly distributed on $(0, 1)$. Note that inequality (7.1), (7.2), and (7.4) are all satisfied in this case. The results under the different policies mentioned are summarized here:

(i) Follow the optimal policy, which takes the following form: at stage 1, stop inspection in (n_1, d_1) if and only if $d_1 < d_{n_1}$, with the values of d_{n_1} summarized in Table 7.1. At stage 2, for each (n_1, n_2), stop inspection if and only if $(d_1, d_2) \in S_{n_1}(n_2)$, where $S_{n_1}(n_2)$ follows the specification at the end of §7.3. An illustration of $S_{n_1}(n_2)$, for $n_1 = 10$ and $n_2 = 15$, is given in Figure 7.1. The expected total cost under this optimal policy is 513.1.

(ii) Follow the heuristic policy specified earlier: at stage 1, treating the defective rate at stage 2 as a constant, the optimal policy takes the form of stopping inspection at (n_1, d_1) if and only if $d_1 < d_{n_1}$, with the threshold values d_{n_1} listed in Table 7.2. At stage 2, we follow the optimal threshold policy in Theorem 7.13, which is the same as the stage-2 policy in (i). Under this policy, the expected total cost is 524.5, less than 3% above the optimal cost.

(iii) Treat the defective rates at both stages as constants. Following Theorem 7.15, this results in a policy that inspects all units in stage 1 and inspects 0 units in stage 2; the expected total cost is 562.5, about 10% above the optimal cost.

Example 7.19 Consider a variation of Example 7.18, with $N = 30$; $c_{i1} = 3.5$, $c_{i2} = 7.5$; $c_{r1} = 1.0$, $c_{r2} = 2.0$, $c_{r12} = 3.5$; $EC(X) = 3.5$, $EC(Y_1) =$

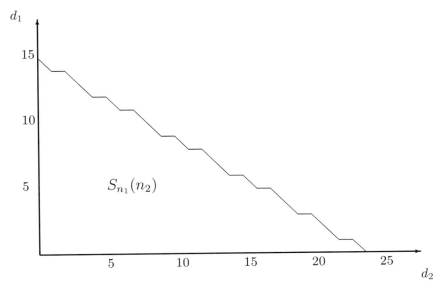

FIGURE 7.1. Threshold structure at stage 2.

13.5, $EC(Y_2) = 15$, $EC(Y_{12}) = 25.5$; and Θ_1 and Θ_2 are both uniformly distributed on $(0, 1)$. The expected total costs are: 357.3 for the optimal policy; 360.5 for the heuristic policy (less than 1% above the optimal cost); and 378.8 for the policy that treats both defective rates as constants (6% above the optimal cost).

From the preceding examples, it should be clear that the optimal policy and the heuristic policy are exactly the same at stage 2: both involve (the same) N^2 threshold curves, such as the one illustrated in Figure 7.1. Each such curve requires no more than N^2 steps to generate. Hence, the overall computational effort is $O(N^4)$ to generate the policy at stage 2. At stage 1, as discussed at the beginning of the section, the optimal policy can be derived via the dynamic programming recursion, with a computational effort of $O(N^2)$; while the heuristic policy can be derived using roughly half of that effort, thanks to its guaranteed threshold structure, with N thresholds.

For implementation, both policies can be precomputed off-line. The inspection is then carried out online: proceed from stage 1 to stage 2; at either stage, inspect one unit at a time, check the number of identified defects against the precomputed threshold values, and stop inspection when the number falls below the threshold. Furthermore, at stage 2, always inspect first those units that have not been inspected at stage 1. Note, however, that for the optimal policy the threshold structure at stage 1 is *not* guaranteed. Hence, in those cases where the optimal policy does not have a threshold structure at stage 1, the heuristic policy has the additional advantage of being easier to implement.

7.6   General Cost Functions

Our results developed thus far are readily extended to a more general cost function (corresponding to warranties or service contracts). For example, as in Chapter 3, we can allow $C(T)$ to be the (warranty) cost function for the whole batch, with T denoting the total (cumulative) lifetime of all N units in the batch. For example,

$$C(T) = (cN)[NW - T]^+/(NW) = c[N - T/W]^+,$$

where $c > 0$ is the selling price of each unit, W is the guaranteed total lifetime of the N units, and $[x]^+$ denotes $\max\{x, 0\}$. In this "cumulative warranty" model ([9, 87]), the manufacturer pays back part of the selling price on a pro rata basis. This type of warranty applies mostly to reliability systems, where spare parts (in cold standby) are used extensively.

We shall focus on stage 2 and show that the structure of the optimal policy in §7.3 remains valid under the general cost function as described earlier. As to stage 1, the comments at the beginning of §7.5 still apply.

Consider the optimal policy at stage 2. As in §7.3, suppose the batch has already been inspected at stage 1, with the inspected items identified as set **a**. Recall that, the lifetimes of the items follow the distributions of X, Y_1, Y_2, and Y_{12}, respectively, for good items and items that are stage-1, stage-2, and 1-2 defects. Assume $X \geq_{st} Y_j \geq_{st} Y_{12}$ for $j = 1, 2$, all lifetimes are independent, and $C(T)$ is a convex (and decreasing) function of T.

Let $Z(\theta_1, \theta_2)$ denote the random variable that is equal to X, Y_1, Y_2, and Y_{12} with probability $(1 - \theta_1)(1 - \theta_2)$, $\theta_1(1 - \theta_2)$, $(1 - \theta_1)\theta_2$, and $\theta_1\theta_2$, respectively. That is, $Z(\theta_1, \theta_2)$ is the lifetime of a unit that is not inspected at either stage, given the defective rates at the two stages are θ_1 and θ_2.

For convenience in notation, write

$$Z_{n,m}(\theta_1, \theta_2) := \sum_{i=n}^{m} Z_i(\theta_1, \theta_2), \quad X_{n,m} := \sum_{i=n}^{m} X_i,$$

where X_i are i.i.d. samples of X, and $Z_i(\theta_1, \theta_2)$ are i.i.d. samples of $Z(\theta_1, \theta_2)$.

In the same spirit as (7.1) and (7.2), we assume that

$$\mathsf{E}C(X_{1,N-1} + Y_1) - \mathsf{E}C(X_{1,N}) \geq c_{r1}, \tag{7.45}$$
$$\mathsf{E}C(X_{1,N-1} + Y_2) - \mathsf{E}C(X_{1,N}) \geq c_{r2}, \tag{7.46}$$

and

$$\mathsf{E}C(X_{1,N-1} + Y_{12}) - \mathsf{E}C(X_{1,N-1} + Y_1) \geq c_{r12} - c_{r1}, \tag{7.47}$$
$$\mathsf{E}C(X_{1,N-1} + Y_{12}) - \mathsf{E}C(X_{1,N-1} + Y_2) \geq c_{r12} - c_{r2}. \tag{7.48}$$

As before, let **a** denote the set of items that have been inspected at stage 1, and let $\bar{\mathbf{a}}$ be its complement. Suppose $|\mathbf{a}| = n_1$, and suppose stage 2 has inspected n_2 items from **a** and $\bar{n}_2$ items from $\bar{\mathbf{a}}$. If stage 2 stops inspection at this point and given $\Theta_1 = \theta_1$ and $\Theta_2 = \theta_2$, then the expected warranty cost is

$$\phi(n_2, \bar{n}_2, \theta_1, \theta_2)$$
$$= \mathsf{E}[C(X_{1,n_2+\bar{n}_2} + Z_{n_2+\bar{n}_2+1,n_1+\bar{n}_2}(0, \theta_2) + Z_{n_1+\bar{n}_2+1,N}(\theta_1, \theta_2)].$$

We first examine the priority structure in Theorem 7.7. As in Section §7.3.1, let $V(n_2, \bar{n}_2, d_1, d_2)$ denote the optimal expected future cost, including inspection and rework cost and the total warranty cost, starting from the state $(n_2, \bar{n}_2, d_1, d_2)$. Similar to (7.14), we need to show

$$\mathsf{E}[V(n_2 + 1, \bar{n}_2, d_1, d_2 + I_2')] - \mathsf{E}[V(n_2, \bar{n}_2 + 1, d_1 + I_1, d_2 + I_2)]$$
$$\geq \mathsf{E}[r(\Theta_1, \Theta_2)|D_1(n_1 + \bar{n}_2) = d_1, D_2(n_2 + \bar{n}_2) = d_2]$$
$$-\mathsf{E}[r(0, \Theta_2)|D_2(n_2 + \bar{n}_2) = d_2]. \tag{7.49}$$

The only modification needed in the proof concerns the inequalities involved in the initial step of the induction (when $n_2 + \bar{n}_2 = N - 2$) and in the step of $B - b \geq \Delta$. Because the two modifications are similar, we illustrate only the latter. The inequality in question here (omitting the conditioning) is

$$\phi(n_2 + 1, \bar{n}_2, \theta_1, \theta_2) - \phi(n_2, \bar{n}_2 + 1, \theta_1, \theta_2)$$
$$\geq r(\theta_1, \theta_2) - r(0, \theta_2).$$

Let

$$\eta := X_{1,n_2+\bar{n}_2+1} + Z_{n_2+\bar{n}_2+2,n_1+\bar{n}_2}(0, \theta_2) + Z_{n_1+\bar{n}_2+2,N}(\theta_1, \theta_2).$$

Then the left-hand side of the last inequality

$$\phi(n_2 + 1, \bar{n}_2, \theta_1, \theta_2) - \phi(n_2, \bar{n}_2 + 1, \theta_1, \theta_2)$$
$$= \mathsf{E}C(\eta + Z(\theta_1, \theta_2)) - \mathsf{E}C(\eta + Z(0, \theta_2))$$
$$= \theta_1(1 - \theta_2)[\mathsf{E}C(\eta + Y_1) - \mathsf{E}C(\eta + X)]$$
$$+\theta_1\theta_2[\mathsf{E}C(\eta + Y_{12}) - \mathsf{E}C(\eta + Y_2)]$$
$$\geq \theta_1(1 - \theta_2)[\mathsf{E}C(X_{1,N-1} + Y_1) - \mathsf{E}C(X_{1,N-1} + X)]$$
$$+\theta_1\theta_2[\mathsf{E}C(X_{1,N-1} + Y_{12}) - \mathsf{E}C(X_{1,N-1} + Y_2)]$$
$$\geq c_{r1}\theta_1(1 - \theta_2) + (c_{r12} - c_{r2})\theta_1\theta_2$$
$$= r(\theta_1, \theta_2) - r(0, \theta_2).$$

Note that the first inequality follows from the decreasing convexity of the function $C(\cdot)$, taking into account $\eta \leq_{\text{st}} X_{1,N-1}$ (also see the proof of Lemma 7.20). The second inequality follows from (7.46) and (7.48).

Therefore, we can follow the state definition of §7.3.2. In particular, if $\Theta_1 = \theta_1$, $\Theta_2 = \theta_2$, and we stop inspection after inspecting n_2 units at stage 2, then the expected warranty cost is

$$\phi(n_2, \theta_1, \theta_2)$$
$$:=\quad EC(X_{1,n_2} + Z_{n_2+1,n_1+n_2}(0, \theta_2) + Z_{n_1+n_2+1,N}(\theta_1, \theta_2)) \quad (7.50)$$

for $n_2 \leq N - n_1 - 1$ and

$$\phi(n_2, \theta_1, \theta_2) := EC(X_{1,n_2} + Z_{n_2+1,N}(0, \theta_2)) \qquad (7.51)$$

for $n_2 \geq N - n_1$.

Substituting $\phi(\cdot)$ of (7.50) and (7.51) into (7.20) yield the problem formulation under the general cost function.

Lemma 7.20 Suppose the (warranty) cost function of the whole batch, $C(\cdot)$, is a decreasing and convex function. Then, given θ_1 and θ_2, $\phi(n_2, \theta_1, \theta_2)$ defined in (7.50) and (7.51) is decreasing and convex in n_2.

Proof. Because θ_1 and θ_2 are given here, write $\phi(n_2, \theta_1, \theta_2)$ as $\phi(n_2)$ for simplicity. Clearly, following the assumed stochastic orderings among the lifetimes, we have

$$X \geq_{st} Z(0, \theta_2) \geq_{st} Z(\theta_1, \theta_2)$$

for any θ_1 and θ_2. Because $C(\cdot)$ is a decreasing function, $\phi(n_2)$ is decreasing in n_2, following the expressions in (7.50) and (7.51).

To prove convexity, we use coupling. Consider first the case of $n_2 \leq N - n_1 - 1$. Because $X \geq_{st} Z(\theta_1, \theta_2)$, we can have, for $j = 1, 2$, $X^j \geq Z^j$ almost surely (a.s.), with X^j and Z^j equal in distribution to X and $Z(\theta_1, \theta_2)$, respectively(while maintaining the independence between X^1 and X^2 and between Z^1 and Z^2). Let τ be the sum of $n_2 - 1$ independent samples of X, n_1 independent samples of $Z(0, \theta_2)$, and $N - n_1 - n_2 - 1$ independent samples of $Z(\theta_1, \theta_2)$. Because $C(\cdot)$ is convex, we have

$$C(\tau + X^1 + X^2) + C(\tau + Z^1 + Z^2)$$
$$\geq \quad C(\tau + X^1 + Z^2) + C(\tau + X^2 + Z^1) \quad \text{a.s.}$$

Taking expectations on both sides yields

$$\phi(n_2 + 1) + \phi(n_2 - 1) \geq \phi(n_2) + \phi(n_2) = 2\phi(n_2),$$

which is the desired convexity.

The case of $n_2 \geq N - n_1$ is similar, with the following modifications: let Z^j, for $j = 1, 2$, be independent samples of $Z(0, \theta_2)$; let τ be the sum of $n_2 - 1$ independent samples of X, and $N - n_2 - 1$ independent samples of $Z(0, \theta_2)$. $\square$

Lemma 7.21 Lemma 7.9 is still valid in the general case when the cost function $C(\cdot)$ is decreasing and convex. Specifically, both (7.24) and (7.25) are decreasing in d_1 and d_2 and increasing in n_2.

Proof. For $n_2 \leq N - n_1 - 1$, let

$$G_1(\theta_1, \theta_2) := \phi(n_2 + 1, \theta_1, \theta_2) - \phi(n_2, \theta_1, \theta_2) + r(\theta_1, \theta_2).$$

We want to show that $G_1(\theta_1, \theta_2)$ is decreasing in θ_1 and θ_2. Same as before, denote

$$\eta := X_{1,n_2} + Z_{n_2+1,n_1+n_2}(0, \theta_2) + Z_{n_1+n_2+2,N}(\theta_1, \theta_2).$$

Note that we have left out the (n_1+n_2+1)th item, for conditioning. Hence,

$$
\begin{aligned}
G_1(\theta_1, \theta_2) &= \mathsf{E}C(\eta + X) - \mathsf{E}C(\eta + Z(\theta_1, \theta_2)) + r(\theta_1, \theta_2) \\
&= \theta_1(1 - \theta_2)[\mathsf{E}C(\eta + X) - \mathsf{E}C(\eta + Y_1)] \\
&\quad + (1 - \theta_1)\theta_2[\mathsf{E}C(\eta + X) - \mathsf{E}C(\eta + Y_2)] \\
&\quad + \theta_1\theta_2[\mathsf{E}C(\eta + X) - \mathsf{E}C(\eta + Y_{12})] + r(\theta_1, \theta_2) \\
&= c_{i2} + \theta_1(1 - \theta_2)[\mathsf{E}C(\eta + X) - \mathsf{E}C(\eta + Y_1) + c_{r1}] \\
&\quad + \theta_2[\mathsf{E}C(\eta + X) - \mathsf{E}C(\eta + Y_2) + c_{r2}] \\
&\quad + \theta_1\theta_2[\mathsf{E}C(\eta + Y_2) - \mathsf{E}C(\eta + Y_{12}) + (c_{r12} - c_{r2})].
\end{aligned}
$$

Because $C(\cdot)$ is decreasing and convex, comparing the expressions in the first and third square brackets with (7.46) and (7.48), we know that they are all nonpositive. Also, if θ_1 is replaced by a larger θ_1', η will further decrease, causing the quantities in the two square brackets in question to decrease (i.e, become more negative). Thus we obtain the desired decreasing property in θ_1. For decreasingness in θ_2, we can follow the same reasoning, reorganizing the derivation as follows:

$$
\begin{aligned}
G_1(\theta_1, \theta_2) &= c_{i2} + \theta_2(1 - \theta_1)[\mathsf{E}C(\eta + X) - \mathsf{E}C(\eta + Y_2) \\
&\quad + \theta_1[\mathsf{E}C(\eta + X) - \mathsf{E}C(\eta + Y_1) + c_{r1}] \\
&\quad + \theta_1\theta_2[\mathsf{E}C(\eta + Y_1) - \mathsf{E}C(\eta + Y_{12}) + (c_{r12} - c_{r1})].
\end{aligned}
$$

Similarly, for $n_2 \geq N - n_1$, let

$$G_0(\theta_1, \theta_2) := \phi(n_2 + 1, \theta_1, \theta_2) - \phi(n_2, \theta_1, \theta_2) + r(0, \theta_2)$$

and denote

$$\xi := X_{1,n_2} + Z_{n_2+2,N}(0, \theta_2).$$

Then

$$
\begin{aligned}
G_0(\theta_1, \theta_2) &= \mathsf{E}C(\xi + X) - \mathsf{E}C(\xi + Z(0, \theta_2)) + r(0, \theta_2) \\
&= c_{i2} + \theta_2[\mathsf{E}C(\xi + X) - \mathsf{E}C(\xi + Y_2) + c_{r2}]
\end{aligned}
$$

is independent of (and hence trivially decreasing in) θ_1 and decreasing in θ_2.

Note that both G_1 and G_0 are increasing in n_2, following Lemma 7.20. Therefore, following the same argument as in the proof of Lemma 7.9 proves the desired decreasing property in d_1 and d_2 and the increasing property in n_2 of (7.24) and (7.25) in the context here. $\square$

Here, As in the earlier case, to prove results such as those in Theorem 7.11, the key is to show that $\Psi_{n_2}(d_1, d_2) - \Phi_{n_2}(d_1, d_2)$ is decreasing d_1 and d_2 and increasing in n_2. We already established these properties in Lemma 7.21. Hence, the rest is to follow the induction steps of Theorem 7.11 (also see the proof of Theorem 7.4).

To summarize, we have the following.

Theorem 7.22 The optimal inspection policy at stage 2 stated in Theorem 7.13 is still optimal under the general cost function, $C(\cdot)$, provided it is decreasing and convex. In particular, Theorem 7.11 still holds under the general cost function, $C(\cdot)$, provided $C(\cdot)$ is a decreasing and convex function.

To conclude, we have seen that coordination is the key in quality control of the two-stage system studied here. The optimal policy at the first stage has to take into account execution at the second stage, in particular, of what can and will be optimally carried out. The second stage then continues with what has been done in the first stage; in particular, it gives priority to those units not yet inspected at the first stage. In addition, a threshold structure has been identified with the optimal policies at both stages.

The two properties, K-submodularity and convexity, are central to the optimality of the threshold policies at both stages. A heuristic policy, which, for deriving stage-1 policy, treats the stage-2 defective rate as constant, appears to generate near-optimal results. On the other hand, treating the defective rates at both stages as constants simplifies the derivation of the optimal policy, at the price of a more substantial departure from optimality. Both the optimal and the heuristic policies can be precomputed off-line. The computational effort is $O(N^2)$ for stage 1 and $O(N^4)$ for stage 2.

7.7   Notes

In many production systems the inspection can be represented as a two-stage scheme. For example, in the context of semiconductor wafer fabrication, the first stage corresponds to a wafer probe, an important step that involves computer-controlled automatic testing, and the second stage involves a more routine, final inspection; see Chapter 6, and Lee [52] and Walker [101]. This chapter is based on Yao and Zheng [107]. Other related recent studies include Cassandras [13] on "yield learning" and Yao and Zheng [108] on process control using Markov decision programming.

In the literature, most studies on quality control in multistage systems assume that the defective rate is a known constant. For example, Lindsay and Bishop [55], assuming a constant defective rate at each stage, and with defective units disposed at a cost, showed that the optimal policy is either 100% or 0% inspection at each stage. Eppen and Hurst [33] generalized the model of [55] to allow imperfect inspection. Ballou and Pazer [3] further discussed the impact of inspector fallibility on the optimal policy under imperfect inspection.

The assumption of a constant defective rate, as pointed out in §3.1, leads to a binomial distribution of the number of defective units in the batch. As the batch size becomes large, the coefficient of variation of the number of defective units can be arbitrarily small, making the model inappropriate in some applications; refer to, e.g., Albin and Friedman [1]. Furthermore, the constant defective rate leads to a static parametric optimization problem, rather than a dynamic control problem, because there is no update in terms of the quality information of the batch as inspection progresses.

8

Optimal Inspection in an Assembly System

In contrast to Chapter 7, where the problem is to coordinate the inspections at two stages in series, here we switch to a *parallel* configuration–an assembly system–and focus on how to coordinate the inspections at component subassemblies.

Specifically, suppose each end-product consists of two different component subassemblies and that quality control of the components is carried out before the final assembly. The problem is to develop an inspection policy on the two component batches so that the total cost is minimized. Although the optimal policy can be derived from solving a dynamic programming problem, there is no guarantee that this will result in a simple control policy, such as the threshold-type sequential inspection policies that were proven optimal in earlier chapters (in particular, Chapters 3 and 6). In fact, as the inspection decisions for the two components are interleaved together, the optimal policy in general will switch back and forth between the two components, destroying any threshold structure. Our focus here is on a class of easily implementable policies that have a simple "single-switch" structure. We show that this class of policies is optimal for a special case of the original problem when one of the two components has a constant defective rate. Furthermore, we illustrate through numerical examples that such policies have near-optimal performance when applied to the original problem.

In §8.1 we describe the model in detail and derive certain basic properties of it. A dynamic programming formulation is presented in §8.2. In §8.3 we identify an optimal policy that has a threshold structure for the case when one of the components has a constant defective rate. Based on this result, in

§8.4 we develop a heuristic inspection policy for the original problem, and illustrate its near-optimal performance through two numerical examples.

8.1   A Two-Component Assembly Model

Suppose two components, termed component 1 and component 2, are produced in batches with N units per batch for both components. They are then assembled into N end-products, which, in turn, are supplied to customers under some types of warranty (or service contract). Suppose the defective rates of the two components are Θ_1 and Θ_2, respectively. Here, $\Theta_i \in [0,1]$ is itself a random variable with a known distribution, representing the defective rate of component i. This works as follows: Θ_i is first sampled from its distribution. Suppose the sample value is θ_i; then each unit in the batch is defective with probability θ_i. The distribution of Θ_i, which is common for all the units in the batch, naturally captures the statistical dependence among the units. In this paper Θ_1 and Θ_2 are supposed to be independent, and we assume that the assembly procedure itself does not produce defective products, i.e. as long as the two components are both nondefective, they will be assembled into a nondefective end-product.

For $i = 1, 2$, suppose the lifetime of a defective component i is Y_i, and a nondefective component i has a lifetime X_i. Assume $X_i \geq_{st} Y_i$ and the lifetimes of the components are independent of one another. If the lifetimes of the two components are Z_1 and Z_2, respectively, then the lifetime of the end-product is assumed to be $Z_1 \wedge Z_2 := \min(Z_1, Z_2)$. Let $Z_i(\theta_i)$ denote the random variable equal in distribution to Y_i(resp. X_i) with probability θ_i(resp. $1 - \theta_i$).

Defectives of both components can be detected by inspection. Suppose the inspection is perfect (i.e., a component is identified as defective if and only if it is defective), and the per-unit inspection cost is $C_I^{(1)}$ and $C_I^{(2)}$, respectively, for the two components. A defective unit can be corrected via repair or rework (and become a nondefective unit) at a per-unit cost of $C_R^{(1)}$ or $C_R^{(2)}$, respectively, for the two components.

Similar to Chapter 3, we will first consider a cumulative warranty cost function of the end-products. Specifically, if the total lifetime of the N assembled units is t, then the warranty cost is $C(t)$, with $C(t)$ assumed to be a convex and decreasing function of t. When $C(t)$ is an additive function, we have the individual warranty case in which the warranty applies to each individual unit. To provide adequate incentive to the repair of any defective unit, we assume

$$\mathsf{E}C(\sum_{j=1}^{N-1} X_{1j} \wedge X_{2j} + Y_1 \wedge Y_2) - \mathsf{E}C(\sum_{j=1}^{N-1} X_{1j} \wedge X_{2j} + X_1 \wedge Y_2)$$

$$\geq C_R^{(1)} \tag{8.1}$$

and

$$
\mathsf{E}C(\sum_{j=1}^{N-1} X_{1j} \wedge X_{2j} + Y_1 \wedge Y_2) - \mathsf{E}C(\sum_{j=1}^{N-1} X_{1j} \wedge X_{2j} + Y_1 \wedge X_2)
$$
$$
\geq C_R^{(2)}. \tag{8.2}
$$

In these inequalities, X_{1j} and X_{2j} are i.i.d. replicas of X_1 and X_2, respectively. (We will apply the same notation to component lifetimes, Y and Z.)

Our problem here is to design a procedure to inspect the components to minimize the total (expected) cost for inspection, possible repair work, and the warranty.

We first derive some basic results for the model.

Lemma 8.1 For any given constant $\tau \geq 0$, the expectation

$$
\mathsf{E}[C(\tau + X_1 \wedge X_2) + C(\tau + Y_1 \wedge Y_2)
$$
$$
- C(\tau + X_1 \wedge Y_2) - C(\tau + Y_1 \wedge X_2)] \leq 0
$$

and is increasing in τ.

Proof. Let

$$
\overline{F}(t) := \overline{F}_{X_1}(t)\overline{F}_{X_2}(t) := \mathsf{P}[X_1 \geq t]\mathsf{P}[X_2 \geq t].
$$

Noticing that

$$
\mathsf{E}[C(\tau + X_1 \wedge X_2)
$$
$$
= -\int_0^\infty C(\tau + t) d\overline{F}(t)
$$
$$
= C(\tau) + \int_0^\infty \overline{F}_{X_1}(t)\overline{F}_{X_2}(t) dC(\tau + t),
$$

we have

$$
\mathsf{E}[C(\tau + X_1 \wedge X_2) + C(\tau + Y_1 \wedge Y_2)
$$
$$
- C(\tau + X_1 \wedge Y_2) - C(\tau + Y_1 \wedge X_2)]
$$
$$
= \int_0^\infty [\overline{F}_{X_1}(t)\overline{F}_{X_2}(t) + \overline{F}_{Y_1}(t)\overline{F}_{Y_2}(t)
$$
$$
- \overline{F}_{X_1}(t)\overline{F}_{Y_2}(t) - \overline{F}_{Y_1}(t)\overline{F}_{X_2}(t)] dC(\tau + t). \tag{8.3}
$$

Note in the integral that $C(\cdot)$ is decreasing and the expression in the square brackets is nonnegative, which follows from

$$
\overline{F}_{X_1}(t) \geq \overline{F}_{Y_1}(t) \quad \text{and} \quad \overline{F}_{X_2}(t) \geq \overline{F}_{Y_2}(t)
$$

for all t. Hence, the integral in (8.3) is nonpositive. The desired increasing-ness in τ follows from the convexity of $C(\cdot)$. $\square$

By conditioning on the value of

$$\sum_{j=1}^{N-1} X_{1j} \wedge X_{2j},$$

an immediate consequence of Lemma 8.1 is that from (8.1) and (8.2), we have

$$\mathsf{E}C(\sum_{j=1}^{N-1} X_{1j} \wedge X_{2j} + Y_1 \wedge X_2) - \mathsf{E}C(\sum_{j=1}^{N-1} X_{1j} \wedge X_{2j} + X_1 \wedge X_2)$$

$$\geq C_R^{(1)} \tag{8.4}$$

and

$$\mathsf{E}C(\sum_{j=1}^{N-1} X_{1j} \wedge X_{2j} + X_1 \wedge Y_2) - \mathsf{E}C(\sum_{j=1}^{N-1} X_{1j} \wedge X_{2j} + X_1 \wedge X_2)$$

$$\geq C_R^{(2)}. \tag{8.5}$$

This is because the right-hand sides of (8.4) and (8.5) dominate the right-hand sides of (8.1) and (8.2), respectively.

Lemma 8.2 It is always better, in terms of reducing the warranty cost, to match confirmed (via inspection) nondefective units when doing the assembly.

Proof. We prove this by an interchange argument. Suppose there is an end-product consisting of a confirmed nondefective component 1 and an uninspected component 2, while another end-product consisting of an uninspected component 1 and a confirmed nondefective component 2. We will show that it is better to assemble the two nondefective components, and the two uninspected components. Let τ denote the total lifetime of the other $N-2$ end-products in the batch. Then it suffices to show the following:

$$\mathsf{E}C(\tau + X_1 \wedge X_2 + Z_1 \wedge Z_2) \leq \mathsf{E}C(\tau + X_1 \wedge Z_2 + Z_1 \wedge X_2), \tag{8.6}$$

where Z_1 and Z_2 are the lifetimes of the uninspected components 1 and 2, respectively. Note that

$$\min[x_1, x_2] + \min[y_1, y_2] \geq \min[x_1, y_2] + \min[y_1, x_2]$$

for all real numbers with $x_1 \geq y_1, x_2 \geq y_2$. Hence, based on $X_i \geq_{st} Z_i$, for $i = 1, 2$, and a standard coupling argument, we can conclude that the argument of the $C(\cdot)$ function on the left side of (8.6) dominates, stochas-tically, the argument of $C(\cdot)$ on the right side. Because $C(\cdot)$ is decreasing, (8.6) must hold. $\square$

8.2   Dynamic Programming Formulation

Let $\phi(n_1, \theta_1, n_2, \theta_2)$ denote the minimal expected warranty cost, given $\Theta_1 = \theta_1$, $\Theta_2 = \theta_2$, and exactly n_1 and n_2 items of components 1 and 2 are inspected. Then, from Lemma 8.2, we have

$$\phi(n_1, \theta_1, n_2, \theta_2)$$
$$= \quad \mathsf{EC}(\sum_{i=1}^{n_1} X_{1i} \wedge X_{2i} + \sum_{i=n_1+1}^{n_2} Z_{1i}(\theta_1) \wedge X_{2i}$$
$$+ \sum_{i=n_2+1}^{N} Z_{1i}(\theta_1) \wedge Z_{2i}(\theta_2)),$$

when $n_1 < n_2$; and

$$\phi(n_1, \theta_1, n_2, \theta_2)$$
$$= \quad \mathsf{EC}(\sum_{i=1}^{n_2} X_{1i} \wedge X_{2i} + \sum_{i=n_2+1}^{n_1} X_{1i} \wedge Z_{2i}(\theta_1)$$
$$+ \sum_{i=n_1+1}^{N} Z_{1i}(\theta_1) \wedge Z_{2i}(\theta_2))$$

when $n_1 \geq n_2$.

To identify the optimal inspection policy, we formulate a dynamic programming problem as follows: denote (n_1, d_1, n_2, d_2) as the state, with n_j as the number of inspected units in component j and d_j as the number of defective units identified from those inspected from component j, $j = 1, 2$. Note that when $n_1, n_2 < N$, there are three control actions available at (n_1, d_1, n_2, d_2): to stop inspection, to continue inspection with component 1, and to continue inspection with component 2. (And when $n_j = N$, the action of inspecting component j is not available, for $j = 1, 2$.) Let $V(n_1, d_1, n_2, d_2)$ denote the optimal expected cost-to-go, starting from the state (n_1, d_1, n_2, d_2). Let $\Phi(n_1, d_1, n_2, d_2)$, $\Psi_1(n_1, d_1, n_2, d_2)$ and $\Psi_2(n_1, d_1, n_2, d_2)$ denote the costs-to-go, starting from (n_1, d_1, n_2, d_2), respectively, for the actions: stop inspection, continue inspection with component 1, and continue inspection with component 2.

Then, for $0 \leq n_1, n_2 < N$, we have

$$\Phi(n_1, d_1, n_2, d_2) \qquad\qquad (8.7)$$
$$= \quad \mathsf{E}[\phi(n_1, \Theta_1, n_2, \Theta_2) | D_1(n_1) = d_1, D_2(n_2) = d_2],$$

$$\Psi_1(n_1, d_1, n_2, d_2) \qquad\qquad (8.8)$$
$$= \quad C_I^{(1)} + C_R^{(1)} \bar{\theta}_1(n_1, d_1) + \mathsf{E}[V(n_1 + 1, d_1 + I_1(n_1, d_1), n_2, d_2)],$$

$$\Psi_2(n_1, d_1, n_2, d_2) \tag{8.9}$$
$$= C_I^{(2)} + C_R^{(2)} \bar{\theta}_2(n_2, d_2) + \mathsf{E}[V(n_1, d_1, n_2 + 1, d_2 + I_2(n_2, d_2))],$$

and

$$V(n_1, d_1, n_2, d_2) \tag{8.10}$$
$$= \min\{\Phi(n_1, d_1, n_2, d_2), \Psi_1(n_1, d_1, n_2, d_2), \Psi_2(n_1, d_1, n_2, d_2)\},$$

$$V(N, d_1, n_2, d_2) = \min\{\Phi(N, d_1, n_2, d_2), \Psi_2(N, d_1, n_2, d_2)\}, \tag{8.11}$$

$$V(n_1, d_1, N, d_2) = \min\{\Phi(n_1, d_1, N, d_2), \Psi_1(n_1, d_1, N, d_2)\} \tag{8.12}$$

with

$$V(N, d_1, N, d_2) = \Phi(N, d_1, N, d_2).$$

Here and in the followings, $D_j(n)$ denotes the (random) number of defective units identified from inspecting n units of component j,

$$\bar{\theta}_j(n_j, d_j) := \mathsf{E}[\Theta_j | D_j(n_j) = d_j],$$

and $I_j(n_j, d_j)$ is a binary (0-1) random variable that equals 1 with probability $\bar{\theta}_j(n_j, d_j)$, for $j = 1, 2$. An optimal policy is one that prescribes actions (in each state) that minimize the right-hand sides of equations (8.10) through (8.12).

Because the inspection decisions for the two components are interleaved together, there is no simple threshold-type optimal policies in general. We will consider a special case: component 2 has a *constant* (i.e., deterministic) defective rate, and derive the optimal policy, which, in turn, will lead to a heuristic policy for the original problem (i.e., component 2, like component 1, has a random defective rate).

For ease of discussion, we shall focus on the case of individual warranty costs. In this case a warranty cost of $C(Z_1 \wedge Z_2)$ is associated with each individual end-product that consists of two components with lifetimes Z_1 and Z_2. Note that in the place of Lemma 8.1, what we need in this case is

$$\mathsf{E}[C(X_1 \wedge X_2) + C(Y_1 \wedge Y_2) - C(X_1 \wedge Y_2) - C(Y_1 \wedge X_2)] \leq 0, \tag{8.13}$$

which only requires $C(\cdot)$ to be a decreasing function: convexity is not needed. As before, the inequality in (8.13) implies that it is always better to match confirmed nondefective components when doing assembly. Also, to provide incentive for repairing any identified defective component, we assume

$$C_R^{(1)} \leq \mathsf{E}[C(Y_1 \wedge Y_2) - C(X_1 \wedge Y_2)], \tag{8.14}$$

$$C_R^{(2)} \leq \mathsf{E}[C(Y_1 \wedge Y_2) - C(Y_1 \wedge X_2)]. \tag{8.15}$$

These, together with (8.13), imply that

$$C_R^{(1)} \leq \mathsf{E}[C(Y_1 \wedge X_2) - C(X_1 \wedge X_2)] \tag{8.16}$$

and

$$C_R^{(2)} \leq \mathsf{E}[C(X_1 \wedge Y_2) - C(X_1 \wedge X_2)]. \tag{8.17}$$

Note that in this case ϕ is a linear function of n_1 and n_2, and can be expressed as follows

$$
\begin{aligned}
&\phi(n_1, \theta_1, n_2, \theta_2) \\
&= n_2 \mathsf{E} C(X_1 \wedge X_2) + (n_1 - n_2) \mathsf{E} C(X_1 \wedge Z_2(\theta_2)) \\
&\quad + (N - n_1) \mathsf{E} C(Z_1(\theta_1) \wedge Z_2(\theta_2))
\end{aligned}
$$

when $n_1 \geq n_2$ and

$$
\begin{aligned}
&\phi(n_1, \theta_1, n_2, \theta_2) \\
&= n_1 \mathsf{E} C(X_1 \wedge X_2) + (n_2 - n_1) \mathsf{E} C(Z_1(\theta_1) \wedge X_2) \\
&\quad + (N - n_2) \mathsf{E} C(Z_1(\theta_1) \wedge Z_2(\theta_2))
\end{aligned}
$$

when $n_1 < n_2$. For convenience, denote

$$\Delta_1 := \mathsf{E} C(X_1 \wedge Y_2) - \mathsf{E} C(X_1 \wedge X_2), \tag{8.18}$$

$$\Delta_2 := \mathsf{E} C(Y_1 \wedge Y_2) - \mathsf{E} C(Y_1 \wedge X_2). \tag{8.19}$$

Clearly $\Delta_1 \geq 0$ and $\Delta_2 \geq 0$. From (8.13), we also have $\Delta_1 \geq \Delta_2$.

Now suppose component 2 has a constant defective rate, i.e., $\Theta_2 \equiv \theta_2$. In this case, $\bar{\theta}_2(n_2, d_2) \equiv \theta_2$ for any n_2 and d_2, and we can see from (8.7) to (8.12) that d_2 plays no role in the decision. Hence, we can simplify the state of the dynamic program to (n_1, d_1, n_2).

8.3 One Component with a Constant Defective Rate

We start with two key lemmas, spelling out how the inspection of component 2, which is assumed to have a constant defective rate, should be carried out.

Lemma 8.3 If it is strictly better to inspect batch 2 than to inspect batch 1 in state (n_1, d_1, n_2), then it is optimal not to inspect component 1, starting from state (n_1, d_1, n_2).

Proof. It suffices to prove that if

$$\Psi_2(n_1, d_1, n_2) < \Psi_1(n_1, d_1, n_2), \tag{8.20}$$

then

$$\min\{\Phi(n_1, d_1, n_2 + 1), \Psi_2(n_1, d_1, n_2 + 1)\}$$
$$< \quad \Psi_1(n_1, d_1, n_2 + 1) \tag{8.21}$$

for $n_2 < N - 1$ and

$$\Phi(n_1, d_1, n_2 + 1) < \Psi_1(n_1, d_1, n_2 + 1) \tag{8.22}$$

for $n_2 = N - 1$. We shall prove this by contradiction.

For $n_2 < N - 1$, if (8.21) is not true, then

$$V(n_1, d_1, n_2 + 1) = \Psi_1(n_1, d_1, n_2 + 1).$$

From this we have

$$\Psi_2(n_1, d_1, n_2)$$
$$= \quad C_I^{(2)} + C_R^{(2)}\theta_2 + V(n_1, d_1, n_2 + 1)$$
$$= \quad C_I^{(2)} + C_R^{(2)}\theta_2 + \Psi_1(n_1, d_1, n_2 + 1)$$
$$= \quad C_I^{(2)} + C_R^{(2)}\theta_2 + C_I^{(1)} + C_R^{(1)}\theta_1(n_1, d_1)$$
$$\quad + \mathsf{E}[V(n_1 + 1, d_1 + I_1(n_1, d_1), n_2 + 1)].$$

But

$$\Psi_1(n_1, d_1, n_2)$$
$$= \quad C_I^{(1)} + C_R^{(1)}\bar{\theta}_1(n_1, d_1) + \mathsf{E}[V(n_1 + 1, d_1 + I_1(n_1, d_1), n_2)]$$
$$\leq \quad C_I^{(1)} + C_R^{(1)}\bar{\theta}_1(n_1, d_1) + \mathsf{E}[\Psi_2(n_1 + 1, d_1 + I_1(n_1, d_1), n_2)]$$
$$= \quad C_I^{(1)} + C_R^{(1)}\bar{\theta}_1(n_1, d_1) + C_I^{(2)} + C_R^{(2)}\theta_2$$
$$\quad + \mathsf{E}[V(n_1 + 1, d_1 + I_1(n_1, d_1), n_2 + 1)]$$
$$= \quad \Psi_2(n_1, d_1, n_2),$$

which contradicts (8.20).

For $n_2 = N - 1$, if (8.22) is not true, then

$$\Psi_1(n_1, d_1, N) = V(n_1, d_1, N) \leq \Phi(n_1, d_1, N).$$

On the other hand, we can rewrite (8.20) as:

$$\Psi_2(n_1, d_1, n_2)$$
$$= \quad C_I^{(2)} + C_R^{(2)}\theta_2 + V(n_1, d_1, N)$$
$$< \quad \Psi_1(n_1, d_1, n_2)$$
$$= \quad C_I^{(1)} + C_R^{(1)}\bar{\theta}_1(n_1, d_1) + \mathsf{E}[V(n_1 + 1, d_1 + I_1(n_1, d_1), N - 1)].$$

Hence,

$$
\begin{aligned}
& C_I^{(2)} + C_R^{(2)}\theta_2 + \Psi_1(n_1, d_1, N) \\
< \ & C_I^{(1)} + C_R^{(1)}\bar{\theta}_1(n_1, d_1) + \mathsf{E}[V(n_1 + 1, d_1 + I_1(n_1, d_1), N - 1)],
\end{aligned}
$$

or equivalently,

$$
\begin{aligned}
& C_I^{(2)} + C_R^{(2)}\theta_2 + \mathsf{E}[V(n_1 + 1, d_1 + I_1(n_1, d_1), N)] \\
< \ & \mathsf{E}[V(n_1 + 1, d_1 + I_1(n_1, d_1), N - 1)].
\end{aligned}
$$

Because

$$
\begin{aligned}
& \mathsf{E}[V(n_1 + 1, d_1 + I_1(n_1, d_1), N - 1] \\
\leq \ & \mathsf{E}[\Psi_2(n_1 + 1, d_1 + I_1(n_1, d_1), N - 1] \\
= \ & C_I^{(2)} + C_R^{(2)}\theta_2 + \mathsf{E}[V(n_1 + 1, d_1 + I_1(n_1, d_1), N)],
\end{aligned}
$$

we have reached a contradiction. $\square$

Lemma 8.4 Given $\Theta_1 = \theta_1$ and that exactly n_1 units of component 1 are inspected, the optimal inspection rule for component 2 is as follows:

(a) inspect all N units, if $C_I^{(2)} \leq [(1 - \theta_1)\Delta_1 + \theta_1\Delta_2 - C_R^{(2)}]\theta_2$;

(b) inspect 0 units, if $C_I^{(2)} > (\Delta_1 - C_R^{(2)})\theta_2$; and

(c) inspect n_1 units, if

$$
[(1 - \theta_1)\Delta_1 + \theta_1\Delta_2 - C_R^{(2)}]\theta_2 < C_I^{(2)} \leq (\Delta_1 - C_R^{(2)})\theta_2.
$$

Here, Δ_1 and Δ_2 are defined in (8.18).

Proof. When $n_2 < n_1$, we have

$$
\begin{aligned}
& C_I^{(2)} + C_R^{(2)}\theta_2 + \phi(n_1, \theta_1, n_2 + 1) - \phi(n_1, \theta_1, n_2) \\
= \ & C_I^{(2)} + C_R^{(2)}\theta_2 + \mathsf{E}C(X_1 \wedge X_2) - \mathsf{E}C(X_1 \wedge Z_2) \\
= \ & C_I^{(2)} + \theta_2(C_R^{(2)} - \Delta_1);
\end{aligned} \tag{8.23}
$$

When $n_2 \geq n_1$, we have

$$
\begin{aligned}
& C_I^{(2)} + C_R^{(2)}\theta_2 + \phi(n_1, \theta_1, n_2 + 1) - \phi(n_1, \theta_1, n_2) \\
= \ & C_I^{(2)} + C_R^{(2)}\theta_2 + \mathsf{E}C(Z_1 \wedge X_2) - \mathsf{E}C(Z_1 \wedge Z_2) \\
= \ & C_I^{(2)} + \theta_2[C_R^{(2)} - (1 - \theta_1)\Delta_1 - \theta_1\Delta_2].
\end{aligned} \tag{8.24}
$$

Here and in what follows, we write $Z_i := Z_i(\theta_i)$, for $i = 1, 2$. Note that both (8.23) and (8.24) are independent of n_2. If $C_I^{(2)}$ falls into the range in case (a), then both (8.23) and (8.24) are nonpositive, taking into account $\Delta_1 \geq \Delta_2$. This means that it is better to stop inspection after $n_2 + 1$ (instead of n_2) units have been inspected, for any $n_2 < N$. Therefore, all units of component 2 should be inspected.

In the case of (b), both (8.23) and (8.24) are nonnegative. Hence, it is always better to stop inspection at n_2 than at $n_2 + 1$, which implies that it is optimal to inspect no units at all.

In the case of (c), (8.23) is nonpositive and (8.24) is nonnegative. This means that continuing inspection is better than stopping inspection when $n_2 < n_1$, but stopping becomes better if $n_2 \geq n_1$. Therefore, exactly $n_2 = n_1$ units of component 2 should be inspected. □

Let

$$\phi'(n_1, \theta_1) := \min_{0 \leq n_2 \leq N} \{\phi(n_1, \theta_1, n_2) + n_2(C_I^{(2)} + C_R^{(2)}\theta_2)\} \qquad (8.25)$$

denote the expected cost-to-go given $\Theta_1 = \theta_1$, and exactly n_1 units of component 1 have been inspected. Then, from Lemma 8.4, we have

$$\phi'(n_1, \theta_1) = n_1 EC(X_1 \wedge Z_2) + (N - n_1)EC(Z_1 \wedge Z_2) \qquad (8.26)$$

if $C_I^{(2)} > \theta_2(\Delta_1 - C_R^{(2)})$;

$$\phi'(n_1, \theta_1) = n_1 EC(X_1 \wedge X_2) + (N - n_1)EC(Z_1 \wedge X_2)$$
$$+ NC_I^{(2)} + NC_R^{(2)}\theta_2 \qquad (8.27)$$

if $C_I^{(2)} \leq \theta_2[(1 - \theta_1)\Delta_1 + \theta_1\Delta_2 - C_R^{(2)}]$; and

$$\phi'(n_1, \theta_1) = n_1 EC(X_1 \wedge X_2) + (N - n_1)EC(Z_1 \wedge Z_2)$$
$$+ n_1 C_I^{(2)} + n_1 C_R^{(2)}\theta_2 \qquad (8.28)$$

if $\theta_2[(1 - \theta_1)\Delta_1 + \theta_1\Delta_2 - C_R^{(2)}] < C_I^{(2)} \leq \theta_2(\Delta_1 - C_R^{(2)})$.

Denote

$$\Phi'(n_1, d_1) := E[\phi'(n_1, \Theta_1)|D_1(n_1) = d_1]. \qquad (8.29)$$

Note that from the dynamic programming formulation, we may generally have different optimal actions (i.e., in terms of inspecting component 1 or 2 or stopping inspection) in different states and may switch actions many times between the two components before stopping the inspection. Let us now focus on a specific class of policies, which we call a *single-switch* policy: starting from $(0, 0, 0)$, we begin with inspecting component 1; once we decide to switch to inspecting component 2, we will never switch back to inspecting component 1. Note that with a single-switch policy if we stop inspecting component 1 at (n_1, d_1) (i.e., when n_1 units have been inspected, of which d_1 are defective), the expected cost-to-go is $\Phi'(n_1, d_1)$, which can be obtained by inspecting component 2 following the optimal inspection rule in Lemma 8.4, with $\theta_1 = \bar{\theta}_1(n_1, d_1)$, the conditional defective rate of component 1 at (n_1, d_1). Let $\Psi'(n_1, d_1)$ be the expected cost-to-go if we continue to inspect component 1, and let $V'(n_1, d_1)$ be the minimal expected cost-to-go by following a single-switch policy, starting from (n_1, d_1). Then

$$
\begin{aligned}
V'(n_1, d_1) &= \min\{\Phi'(n_1, d_1), \Psi'(n_1, d_1)\} & \text{if } n_1 < N, \\
V'(n_1, d_1) &= \Phi'(n_1, d_1) & \text{if } n_1 = N,
\end{aligned}
$$

and

$$
\begin{aligned}
&\Psi'(n_1, d_1) \\
&= C_I^{(1)} + C_R^{(1)}\theta_1(n_1, d_1) + \mathsf{E}[V'(n_1 + 1, d_1 + I_1(n_1, d_1))] \quad (8.30)
\end{aligned}
$$

if $n_1 < N$.

Theorem 8.5 points out that the optimal inspection policy (for the case of component 2 having a constant defective rate) lies within the class of single-switch policies.

Recall that $\Phi(n_1, d_1, n_2)$, $\Psi_1(n_1, d_1, n_2)$, $\Psi_2(n_1, d_1, n_2)$, and $V(n_1, d_1, n_2)$ denote the expected costs-to-go starting from state (n_1, d_1, n_2), respectively, for stopping inspection, continuing inspection of component 1, continuing inspection of component 2, and following the global optimal policy.

Theorem 8.5 $V'(n_1, d_1) = V(n_1, d_1, 0)$ for all (n_1, d_1). Consequently, it is optimal to follow a single-switch policy that starts with inspecting component 1 and then switches to inspecting component 2 following the rule described in Lemma 8.4 (with $\theta_1 = \bar{\theta}_1(n_1, d_1)$, provided the switching takes place at (n_1, d_1)).

Proof. Obviously, we have $V'(n_1, d_1) \geq V(n_1, d_1, 0)$, following the optimality of V. We will prove

$$V'(n_1, d_1) \leq V(n_1, d_1, 0) \tag{8.31}$$

by induction on n_1. When $n_1 = N$, there is no uninspected component 1, and hence,

$$V(N_1, d_1, 0) = V'(N_1, d_1)$$

holds trivially.

Assuming (8.31) holds true for all $n_1 \geq k + 1$, we now prove it holds for $n_1 = k$. Let $\Phi'(n_1, d_1)$ and $\Psi'(n_1, d_1)$ denote the cost-to-go functions, respectively, for stopping and continuing inspection in the single-switch policy. Clearly, we have $\Phi(n_1, d_1, 0) \geq \Phi'(n_1, d_1)$, because there might be some inspected component 2 included in Φ', whereas there is no such component in Φ. Note that we also have

$$\Psi_1(n_1, d_1, 0)$$
$$= C_I^{(1)} + C_R^{(1)}\theta_1(n_1, d_1) + \mathsf{E}V(n_1 + 1, d_1 + I_1(n_1, d_1), 0)$$
$$\geq C_I^{(1)} + C_R^{(1)}\theta_1(n_1, d_1) + \mathsf{E}V'(n_1 + 1, d_1 + I_1(n_1, d_1))$$
$$= \Psi'(n_1, d_1),$$

where the inequality follows from the induction hypothesis. If

$$\Psi_2(n_1, d_1, 0) \geq \min\{\Phi(n_1, d_1, 0), \Psi_1(n_1, d_1, 0)\}, \tag{8.32}$$

then

$$V(n_1, d_1, 0)$$
$$= \min\{\Phi(n_1, d_1, 0), \Psi_1(n_1, d_1, 0)\}$$
$$\geq \min\{\Phi'(n_1, d_1), \Psi'(n_1, d_1)\}$$
$$= V'(n_1, d_1),$$

which is what is desired. On the other hand, if

$$\Psi_2(n_1, d_1, 0) < \min\{\Phi(n_1, d_1, 0), \Psi_1(n_1, d_1, 0)\},$$

then following Lemma 8.3, starting from $(n_1, d_1, 0)$, we should not inspect any more unit of component 1. Hence,

$$V(n_1, d_1, 0) = \Phi'(n_1, d_1) \geq V'(n_1, d_1). \qquad \square$$

We will develop the optimal inspection policy for component 1, i.e., the policy before switching to inspecting component 2. To this end, we need to show that the ϕ' function in (8.25) satisfies K-submodularity as defined in Chapter 3.

Lemma 8.6 $\phi'(n_1, \theta_1)$ is K-submodular in (n_1, θ_1) with $K = C_R^{(1)}$.

Proof. Suppose $\theta_1 < \theta_1'$. We want to show

$$\phi'(n_1, \theta_1) - \phi'(n_1 + 1, \theta_1) + \phi'(n_1 + 1, \theta_1') - \phi'(n_1, \theta_1') + C_R^{(1)}(\theta_1' - \theta_1)$$
$$\leq 0. \tag{8.33}$$

We shall establish this inequality by considering several cases corresponding to the range of values of $C_I^{(2)}$. For convenience, denote $Z_i' := Z_i(\theta_i')$ for $i = 1, 2$.

If $C_I^{(2)} > \theta_2(\Delta_1 - C_R^{(2)})$, by (8.26) the LHS (left-hand side) of (8.33) becomes

$$\mathsf{E}C(Z_1 \wedge Z_2) - \mathsf{E}C(X_1 \wedge Z_2) + \mathsf{E}C(X_1 \wedge Z_2)$$

$$-\mathsf{E}C(Z_1' \wedge Z_2) + C_R^{(1)}(\theta_1' - \theta_1)$$
$$= \{C_R^{(1)} - [\mathsf{E}C(Y_1 \wedge Z_2) - \mathsf{E}C(X_1 \wedge Z_2)]\}(\theta_1' - \theta_1)$$
$$= \left[\theta_2\{C_R^{(1)} - [\mathsf{E}C(Y_1 \wedge Y_2) - \mathsf{E}C(X_1 \wedge Y_2)]\}\right.$$
$$\left. + (1 - \theta_2)\{C_R^{(1)} - [\mathsf{E}C(Y_1 \wedge X_2) - \mathsf{E}C(X_1 \wedge X_2)]\}\right](\theta_1' - \theta_1)$$
$$\leq 0,$$

where the inequality follows from (8.14 and 8.16).

If $C_I^{(2)} \leq [(1 - \theta_1')\Delta_1 + \theta_1'\Delta_2 - C_R^{(2)}]\theta_2$, because $\Delta_1 \geq \Delta_2$, we have

$$C_I^{(2)} \leq [(1 - \theta_1)\Delta_1 + \theta_1\Delta_2 - C_R^{(2)}]\theta_2,$$

and hence, the LHS of (8.33) is equal to

$$\mathsf{E}C(Z_1 \wedge X_2) - \mathsf{E}C(X_1 \wedge X_2) + \mathsf{E}C(X_1 \wedge X_2)$$
$$-\mathsf{E}C(Z_1' \wedge X_2) + C_R^{(1)}(\theta_1' - \theta_1)$$
$$= \{C_R^{(1)} - [\mathsf{E}C(Y_1 \wedge X_2) - \mathsf{E}C(X_1 \wedge X_2)]\}(\theta_1' - \theta_1)$$
$$\leq 0.$$

Here the inequality again follows from (8.14).

Next, suppose

$$[(1 - \theta_1)\Delta_1 + \theta_1\Delta_2 - C_R^{(2)}]\theta_2 < C_I^{(2)} \leq \theta_2(\Delta_1 - C_R^{(2)}).$$

Note that this implies

$$C_I^{(2)} \geq [(1 - \theta_1')\Delta_1 + \theta_1'\Delta_2 - C_R^{(2)}]\theta_2.$$

The LHS of (8.33) is then equal to

$$\mathsf{E}C(Z_1 \wedge Z_2) - \mathsf{E}C(Z_1' \wedge Z_2) + C_R^{(1)}(\theta_1' - \theta_1)$$
$$= (\theta_1' - \theta_1)\{\theta_2[\mathsf{E}C(X_1 \wedge Y_2) - \mathsf{E}C(Y_1 \wedge Y_2) + C_R^{(1)}]$$
$$+ (1 - \theta_2)[\mathsf{E}C(X_1 \wedge X_2) - \mathsf{E}C(Y_1 \wedge X_2) + C_R^{(1)}]\}$$
$$\leq 0.$$

The last case is

$$[(1 - \theta_1')\Delta_1 + \theta_1'\Delta_2 - C_R^{(2)}]\theta_2 < C_I^{(2)} \leq [(1 - \theta_1)\Delta_1 + \theta_1\Delta_2 - C_R^{(2)}]\theta_2. \quad (8.34)$$

In this case, (8.33) becomes

$$C_I^{(2)} + C_R^{(2)}\theta_2 + \mathsf{E}C(Z_1 \wedge X_2) - \mathsf{E}C(Z_1' \wedge Z_2) + (\theta_1' - \theta_1)C_R^{(1)} \leq 0,$$

which is equivalent to

$$C_I^{(2)} + C_R^{(2)}\theta_2 + \theta_1'\theta_2(\Delta_1 - \Delta_2) - \theta_2\Delta_1$$
$$- (\theta_1' - \theta_1)[\mathsf{E}C(Y_1 \wedge X_2) - \mathsf{E}C(X_1 \wedge X_2) - C_R^{(1)}] \leq 0.$$

With condition (8.34), it suffices to show

$$(\theta_1' - \theta_1)\{\theta_2(\Delta_1 - \Delta_2) - [\mathsf{E}C(Y_1 \wedge X_2) - \mathsf{E}C(X_1 \wedge X_2) - C_R^{(1)}]\} \le 0,$$

or equivalently,

$$(1 - \theta_2)[\mathsf{E}C(Y_1 \wedge X_2) - \mathsf{E}C(X_1 \wedge X_2) - C_R^{(1)}]$$
$$+\theta_2[\mathsf{E}C(Y_1 \wedge Y_2) - \mathsf{E}C(X_1 \wedge Y_2) - C_R^{(1)}]$$
$$\ge \quad 0,$$

which follows from (8.1 and 8.4). $\square$

In view of these results, with the $\Phi'(n_1, d_1)$ and $\Psi'(n_1, d_1)$ defined in (8.29) and (8.30), by exactly the same argument as in Theorems 3.12 and 3.14 of Chapter 3, we can prove the following.

Lemma 8.7 $\Psi'(n_1, d_1) - \Phi'(n_1, d_1)$ is decreasing in d_1 and increasing in n_1.

For $0 \le n_1 \le N$, let

$$d_{1,n_1}^* := \min\{d_1 \le n_1 : \Psi'(n_1, d_1) < \Phi'(n_1, d_1)\}. \tag{8.35}$$

If $\Psi'(n_1, d_1) \ge \Phi'(n_1, d_1)$ for all $d_1 \le n_1$, we define $d_{1,n_1}^* := n_1 + 1$.

Theorem 8.8 (a) It is optimal to stop inspecting component 1 at (n_1, d_1) if and only if $d_1 < d_{1,n_1}^*$.
(b) d_{1,n_1}^* is increasing in n_1, i.e., $d_{1,n_1}^* \le d_{1,n_1+1}^*$.

Proof. Note that it is optimal to stop inspecting component 1 at (n_1, d_1) if and only if

$$\Psi'(n_1, d_1) \ge \Phi'(n_1, d_1),$$

which is equivalent to $d_1 < d_{1,n_1}^*$, because $\Psi'(n_1, d_1) - \Phi'(n_1, d_1)$ is decreasing in d_1 from Lemma 8.7. This proves (a). For (b), note that

$$\Psi'(n_1 + 1, d_1) < \Phi'(n_1 + 1, d_1)$$

implies $\Psi'(n_1, d_1) < \Phi'(n_1, d_1)$, because $\Psi'(n_1, d_1) - \Phi'(n_1, d_1)$ is increasing in n_1. Hence,

$$\{d_1 \le n_1 : \Psi'(n_1, d_1) < \Phi'(n_1, d_1)\}$$
$$\supseteq \quad \{d_1 \le n_1 : \Psi'(n_1 + 1, d_1) < \Phi'(n_1 + 1, d_1)\},$$

which implies $d_{1,n_1}^* \le d_{1,n_1+1}^*$. $\square$

Note that from (8.29),

$$\Phi'(n_1, d_1) = \mathsf{E}[\phi'(n_1, \Theta_1)|D_1(n_1) = d_1];$$

and for given θ_1, $\phi'(n_1, \theta_1)$ is calculated through (8.26), (8.27), or (8.28), corresponding to different values of θ_2. Hence, in general, both $\Phi'(n_1, d_1)$ and $\Psi'(n_1, d_1)$, as well as d_{1,n_1}^*, depend on θ_2, the defective rate of component 2.

In summary, we have the following.

Theorem 8.9 In the case when component 2 has a constant defective rate θ_2, the following single-switch policy is optimal:

- first inspect the batch of component 1, then inspect the batch of component 2;

- component 1 is inspected following the threshold rule in Theorem 8.8;

- suppose the inspection of component 1 stops at (n_1, d_1) (i.e., n_1 units have been inspected, of which d_1 units are found to be defective), then component 2 is inspected following the rule in Lemma 8.4, with $\theta_1 = E[\Theta_1 | D_1(n_1) = d_1]$.

Note that by symmetry, if component 1, instead of component 2, has a constant defective rate, we have the same inspection policy, simply switching the role of the two components in preceding discussion.

To conclude this section, we consider the further specialized case when *both* components have constant defective rates, i.e., $\Theta_1 \equiv \theta_1$ and $\Theta_2 \equiv \theta_2$, the problem is much simpler. Because there is no information concerning the quality of the two components as inspection progresses in this case, it becomes a static optimization with decision variables n_1 and n_2, the number of units to inspect from the two components. Note that there is a symmetry among all assembled end-products, such that every product should follow the same (optimal) decision regarding whether to inspect the two components that are assembled to form the product. In other words, all we need is to find the best among the four decisions, for each pair of components (any pair among the N pairs): inspect both components, inspect component 1 only, inspect component 2 only, and inspect neither. These correspond to the following costs:

$$C_I^{(1)} + \theta_1 C_R^{(1)} + C_I^{(2)} + \theta_2 C_R^{(2)} + EC(X_1 \wedge X_2), \quad C_I^{(1)} + \theta_1 C_R^{(1)} + EC(X_1 \wedge Z_2),$$

$$C_I^{(2)} + \theta_2 C_R^{(2)} + EC(Z_1 \wedge X_2), \quad EC(Z_1 \wedge Z_2).$$

(In particular, note that here N plays no role in the decision.)

By comparing the four costs we can easily specify the ranges for the values of θ_1 and θ_2 and the corresponding optimal solution (n_1^*, n_2^*), which are summarized here:

(a) $n_1^* = 0$, $n_2^* = N$ if $\theta_1 < \theta_{1a}^*$ and $\theta_2 \geq \theta_{2b}^*$;

(b) $n_1^* = n_2^* = N$ if $\theta_1 \geq \theta_{1a}^*$ and $\theta_2 \geq \theta_{2b}^*$, or if $\theta_1 \geq \theta_{1c}^*$ and $\theta_{2a}^* \leq \theta_2 < \theta_{2b}^*$;

(c) $n_1^* = n_2^* = 0$ if $\theta_1 < \theta_{1b}^*$ and $\theta_2 < \theta_{2a}^*$, or if $\theta_1 < \theta_{1c}^*$ and $\theta_{2a}^* \leq \theta_2 < \theta_{2b}^*$;

(d) $n_1^* = N$, $n_2^* = 0$ if $\theta_1 \geq \theta_{1b}^*$ and $\theta_2 < \theta_{2a}^*$.

Here

$$\theta_{1a}^* := \frac{C_I^{(1)}}{\Delta_1' - C_R^{(1)}}, \qquad \theta_{1b}^* := \frac{C_I^{(1)}}{(1 - \theta_2)\Delta_1' + \theta_2\Delta_2' - C_R^{(1)}},$$

$$\theta_{1c}^* := \frac{C_I^{(1)} + C_I^{(2)} - \theta_2(\Delta_1 - C_R^{(2)})}{(1 - \theta_2)\Delta_1' + \theta_2\Delta_2' - C_R^{(1)}}, \qquad \theta_{2a}^* := \frac{C_I^{(2)}}{\Delta_1 - C_R^{(2)}},$$

and

$$\theta_{2b}^* := \frac{C_I^{(2)}}{(1 - \theta_1)\Delta_1 + \theta_1\Delta_2 - C_R^{(2)}},$$

with

$$\Delta_1' := \mathsf{E}[C(Y_1 \wedge X_2) - C(X_1 \wedge X_2)], \qquad \Delta_2' := \mathsf{E}[C(Y_1 \wedge Y_2) - C(X_1 \wedge Y_2)],$$

and Δ_1 and Δ_2 defined in (8.18).

8.4   A Heuristic Policy

Here we propose a threshold-type *single-switch* heuristic policy to the original problem (i.e., with both components having random defective rates), based on the optimal policy for the special case (of component 2 having a constant defective rate) in Theorem 8.9.

For each pair of (n_2, d_2), with $n_2 = 0, 1, ..., N$ and $d_2 \leq n_2$, denote $d_{1,n_1}^*(n_2, d_2)$, $n_1 = 0, 1, \cdots, N - 1$, as the thresholds for inspecting component 1 as in Theorem 8.8, by assuming component 2 has constant defective rate $\theta_2 = \mathsf{E}[\Theta_2|D_2(n_2) = d_2]$. Similarly, by switching the role of components 1 and 2, we can define $d_{2,n_2}^*(n_1, d_1)$; $n_2 = 0, 1, \cdots, N - 1$ as the thresholds for inspecting component 2 for each pair of (n_1, d_1), assuming component 1 has constant defective rate $\theta_1 = \mathsf{E}[\Theta_1|D_1(n_1) = d_1]$.

Based on these thresholds, we have the following two policies:

- *Policy I.* Start by inspecting component 1 in state $(0, 0, 0, 0)$, following the single-switch threshold rule in follows:

    (1) inspect component 1 one unit at a time, and switch to inspecting component 2 at $(n_1, d_1, 0, 0)$ if and only if

    $$d_1 < d_{1,n_1}^* := d_{1,n_1}^*(0, 0),$$

    or $n_1 = N$;

(2) inspect component 2 one unit a time and stop inspection in state (n_1, d_1, n_2, d_2) if and only if $d_2 < d^*_{2,n_2}(n_1, d_1)$ or $n_2 = N$.

- *Policy II.* Start by inspecting component 2 in state $(0, 0, 0, 0)$, following the single-switch threshold rule in follows:

  (1) inspect component 2 one unit at a time and switch to inspecting component 1 at $(0, 0, n_2, d_2)$ if and only if

  $$d_2 < d^*_{2,n_2} := d^*_{2,n_2}(0, 0)$$

  or $n_2 = N$;

  (2) inspect component 1 one unit a time and stop inspection at state (n_1, d_1, n_2, d_2) if and only if $d_1 < d^*_{1,n_1}(n_2, d_2)$ or $n_1 = N$.

For a given problem, we can easily calculate the expected cost under each of these two policies. The one with lower cost is chosen as our heuristic inspection policy.

Numerical examples have shown that the heuristic policy performs very well: the difference between the expected cost of the heuristic policy and the optimal expected cost is negligible in all examples we tested.

We will illustrate the performance of this heuristic policy in two examples, and compare it against the optimal policy. In addition, we compare it to the solution obtained by treating the defective rates of both components as constants, i.e., assuming $\Theta_1 \equiv \mathsf{E}[\Theta_1]$ and $\Theta_2 \equiv \mathsf{E}[\Theta_2]$.

Example 8.10 Suppose $N = 30$; $C_I^{(1)} = 6.5$, $C_I^{(2)} = 6.3$; $C_R^{(1)} = 1.0$, $C_R^{(2)} = 2.0$; $\mathsf{E}[C(X_1 \wedge X_2)] = 3.5$, $\mathsf{E}[C(X_1 \wedge Y_2)] = 18.5$, $\mathsf{E}[C(Y_1 \wedge X_2)] = 20.0$, $\mathsf{E}[C(Y_1 \wedge Y_2)] = 33.5$; Θ_1 and Θ_2 are both uniformly distributed on $(0,1)$. Note that both inequalities in (8.13) and (8.14) are satisfied in this case. The results under the different policies mentioned are summarized here:

(i) following the optimal policy, which is derived by solving the dynamic programming in §8.2, the expected total cost is 465.609.

(ii) following the heuristic policy, the expected total cost is 465.614 (corresponding to Policy I, whereas the expected total cost under Policy II is 465.649). Under this policy, we start by inspecting component 1 in state $(0, 0, 0, 0)$, and switch to inspecting component 2 at $(n_1, d_1, 0, 0)$ if and only if $d_1 < d^*_{1,n_1}$ or $n_1 = N$; Suppose we switch to inspecting component 2 at $(n_1, d_1, 0, 0)$; then we stop inspection at (n_1, d_1, n_2, d_2) if and only if $d_2 < d^*_{2,n_2}(n_1, d_1)$ or $n_2 = N$. Here d^*_{1,n_1}, $n_1 = 0, 1, \cdots, N - 1$ are summarized in Table 8.1. In addition, $d^*_{2,n_2}(n_1, d_1)$, $n_2 = 0, 1, \cdots, N - 1$, when $n_1 = 15$ and $d_1 = 5$, for example, are presented in the Table 8.2.

(iii) Suppose we treat the defective rates of both components as constants by assuming $\Theta_1 \equiv 0.5$ and $\Theta_2 \equiv 0.5$. Then following the results derived at the end of the last section, we derive the optimal policy as inspecting all

n_1	0	1	2	3	4	5	6	7	8	9
d^*_{1,n_1}	0	0	0	1	1	2	2	2	3	3
n_1	10	11	12	13	14	15	16	17	18	19
d^*_{1,n_1}	4	4	5	5	5	6	6	7	7	8
n_1	20	21	22	23	24	25	26	27	28	29
d_{1,n_1}	8	9	9	10	10	11	11	12	12	13

TABLE 8.1. Threshold values at component 1.

n_2	0	1	2	3	4	5	6	7	8	9
$d^*_{2,n_2}(15,5)$	0	0	1	1	2	2	3	3	4	4
n_2	10	11	12	13	14	15	16	17	18	19
$d^*_{2,n_2}(15,5)$	5	5	6	6	7	8	8	9	9	7
n_2	20	21	22	23	24	25	26	27	28	29
$d^*_{2,n_2}(15,5)$	10	10	11	12	12	13	13	14	14	15

TABLE 8.2. Threshold values at component 2.

units of the two components, resulting in an expected total cost of 534.00, which is much higher than the cost achieved under the heuristic policy.

Example 8.11 Suppose $N = 30$; $C_I^{(1)} = 3.5$, $C_I^{(2)} = 2.3$; $C_R^{(1)} = 2.0$, $C_R^{(2)} = 2.5$; $E[C(X_1 \wedge X_2)] = 3.5$, $E[C(X_1 \wedge Y_2)] = 18.5$, $E[C(Y_1 \wedge X_2)] = 20.0$, $E[C(Y_1 \wedge Y_2)] = 33.5$; Θ_1 is uniformly distributed on $(0.05, 0.40)$, and Θ_2 is uniformly distributed on $(0.05, 0.50)$.

For this problem, the total costs under the optimal policy, the heuristic policy, and the policy corresponding to treating both defective rates as constants are, respectively, 297.357, 297.360, and 306.00.

8.5 Notes

This chapter is based on Zheng [110]. There are many studies focusing on different aspects of assembly systems, e.g., Glasserman and Wang [38], Song et al [88, 89], and Yao [105]. In the literature of quality control, most studies focus on single-stage or serial systems, e.g., Ballou and Pazer [3], Eppen and Hurst [33], Tapiero and Lee [94], Djamaludin et al [32], and Yao and Zheng [107]. There are also studies considering nonserial systems including assembly lines (e.g., Britney [10] and Gunter and Swanson [43]), but the emphases are usually not on deriving optimal inspection policies.

In the discussions in this chapter, we have focused on individual warranty costs. We expect that the results are also applicable to the more general warranty cost functions discussed in §8.1 (also refer to Chapter 3).

Furthermore, the model studied here can be extended to more than two components. Analogous to the heuristic policy, which treats one of the two components as having a constant defective rate, in the general case we will need to treat $M-1$ components, i.e., all but one component, as having constant defective rates. The main difficulty is to prove the K-submodularity of the $\phi'(\cdot)$ function, which still follows the definition in (8.25) but with n_2 replaced by an $(M-1)$-dimensional vector. Another possible extension is to allow inspection for the end-product. In this case we need to consider the coordination of quality control between the components and the end-products, in addition to the coordination among the components. In this regard, the two-stage coordination model studied in [107] may prove helpful.

9

Coordinated Replenishment and Rework with Unreliable Supply Sources

Having started in Chapter 3 from the customer end and gone through several different configurations of production facilities in Chapters 4 through 8, we have now reached the other end of the supply chain. Our focus here is on a set of suppliers, which have different grades of quality and collectively form the sources of supply to a production-inventory system. The system, in turn, fills in a single stream of customer demands. In addition to placing orders of different size with the difference sources, the system can choose to inspect a certain number of units received from each source and repair any identified defectives. We first study this optimal inspection decision, assuming the order quantities are given. This is formulated as an integer optimization problem, which minimizes an objective function that takes into account the inspection and repair costs, as well as the penalty for unmet demand and the salvage value of any surplus units. As the objective function is nonlinear and nonseparable, such an integer optimization problem is difficult to solve in general. However, due to certain special properties of the objective function—properties that appear to be strengthenings of supermodularity and convexity (Proposition 9.1), the optimal solution to the inspection problem has a special structure: it is only necessary to inspect those units supplied from sources that fall in the middle range of the quality spectrum, i.e., we can forgo inspection of units from the best and worst sources. Consequently, the optimal solution is easily identified through a greedy algorithm.

When the replenishment decision is incorporated into the model, the problem structure essentially remains intact. The optimal solution to the order quantities can also be greedily generated, and inspection is optimally applied only to (at most) a single source identified by the greedy algorithm. Furthermore, in the case of linear cost functions (in terms of quality), it is optimal to place orders from two sources only, i.e., dual sourcing is optimal.

These single-period results can be extended to an infinite horizon, with a long-run average cost objective. The optimal policy is an order-up-to policy, with the order-up-to level derived from solving a single-period problem.

When supply imperfection takes the form of a reduced quantity ("yield loss"), the model is easily adapted to generate the optimal replenishment decisions. In fact, the quality control mechanism can be translated into a provision of paying an additional premium to guarantee the delivery quantity.

We will start with a model description and formulation in §9.1, focusing first on the optimal inspection problem, assuming the replenishment decisions have already been made. Key properties of the objective function are established in §9.2, and the optimal inspection problem is solved in §9.3 via a greedy algorithm. Solutions to the optimal replenishment quantities, taking into account the inspection decision, are studied in §9.4. Extensions to the infinite horizon are presented in §9.5. Adaptation of the model to the setting of yield loss is presented in §9.6.

9.1 The Inspection/Rework Model

There are k sources of supply, indexed by $i = 1, ..., k$; and we shall refer to the products supplied from source i as type i. Each unit from source i has a defective rate of θ_i—it is defective with probability θ_i and nondefective with probability $1 - \theta_i$, independent of all other units. All the defective rates are given constants, with $0 < \theta_1 < \theta_2 < \cdots < \theta_k \leq 1$. In addition, denote $\theta_0 \equiv 0$, signifying a nondefective unit. (As noted in Chapter 3, this simple binomial defect model may not be appropriate in some applications, and a better and more general model is to make θ_i a random variable too. This, however, will result in an added layer of sequential decisions for inspection, which we do not treat here.)

We start assuming the order (batch) size from source i, N_i, as given, for all $i = 1, ..., k$; later in §9.4, the order sizes will be treated as decision variables. There is a random demand, denoted D, with a known distribution. The demand can be supplied by products from all k sources, along with some type of warranty or service contract. Suppose the expected warranty/service costs associated with a defective unit and a nondefective unit are, respectively, C_d and C_g; and denote $\Delta := C_d - C_g$. Naturally, assume $\Delta \geq 0$. Defective units can be identified through inspection. Each identified defective unit will be repaired and converted into a nondefective unit. (Hence, the defective rate of any inspected unit is $\theta_0 = 0$.) The unit inspection and repair costs are C_I and C_R, respectively. To ensure that there is enough incentive to repair all identified defective units, assume $C_R \leq \Delta$.

For any surplus unit after demand is satisfied, there is a salvage value, which is a decreasing function of the defective rate, denoted $s(\theta)$. Hence, the salvage value for a surplus unit of batch i is $s(\theta_i)$ if the unit is not inspected; whereas any surplus unit that is inspected (and repaired if necessary) has a salvage value of $s(0)$.

We further assume that supplying demand from a type with a lower index is less costly:

$$C_d\theta_{i-1} + C_g(1 - \theta_{i-1}) - s(\theta_i) \leq C_d\theta_i + C_g(1 - \theta_i) - s(\theta_{i-1})$$

or

$$\Delta(\theta_i - \theta_{i-1}) \geq s(\theta_{i-1}) - s(\theta_i), \qquad \text{for } i = 1, 2, \cdots, k. \tag{9.1}$$

This guarantees that any demand will always be supplied by the best available unit in terms of quality: starting from the inspected units, followed by the (uninspected) units in batch 1, then batch 2, and so forth.

Let $(n_1, \cdots, n_k)$ denote the "state" variable, in which n_i units of batch i have been inspected, $i = 1, ..., k$, with any identified defectives repaired. For convenience, denote

$$n_{i,j} := \sum_{\ell=i}^{j} n_\ell \quad \text{and} \quad N_{i,j} := \sum_{\ell=i}^{j} N_\ell \qquad \text{for } 1 \leq i \leq j \leq k;$$

and denote $n_{i,j} = N_{i,j} = 0$ if $i > j$. Let $W(D, n_1, \cdots, n_k)$ denote the warranty cost minus salvage value, given that the demand is D and inspection is terminated in state $(n_1, \cdots, n_k)$. Then

$$\begin{aligned}
&W(D, n_1, \cdots, n_k) \\
=\ & \min\{n_{1,k}, D\}C_g \\
&+ \sum_{i=1}^{k} \min\{N_i - n_i, (D - n_{i,k} - N_{1,i-1})^+\}(C_g + \theta_i\Delta) \\
&- (n_{1,k} - D)^+ s(0) \\
&- \sum_{i=1}^{k} [N_i - n_i - (D - n_{i,k} - N_{1,i-1})^+]^+ s(\theta_i). \tag{9.2}
\end{aligned}$$

Note that the first two terms on the right-hand side correspond to the warranty costs for inspected and uninspected units that are used to supply demand, while the other two terms correspond to the salvage value of the surplus units that are inspected and uninspected.

Let $\Pi(n_1, n_2, \cdots, n_k)$ denote the expected total cost—including inspection and repair costs, as well as the warranty cost minus salvage value, if we stop inspection in state $(n_1, \cdots, n_k)$. Then

$$\Pi(n_1, n_2, \cdots, n_k)$$
$$= C_I n_{1,k} + C_R \sum_{i=1}^{k} \theta_i n_i + \mathsf{E}[W(D, n_1, n_2, \cdots, n_k)] \qquad (9.3)$$

for $n_i \le N_i$, $i = 1, \cdots, k$. We want to find the best solution $(n_1^*, n_2^*, \cdots, n_k^*)$, the number of units inspected for each batch, to minimize the expected total cost Π.

9.2   Properties of the Cost Function

The properties of the cost function Π in Proposition 9.1 will play a central role in identifying the structure of the optimal solution.

Proposition 9.1 For $i = 2, ..., k$,

$$\Pi(n_1, \cdots, n_{i-1} + 1, n_i, \cdots, n_k) - \Pi(n_1, \cdots, n_{i-1}, n_i + 1, \cdots, n_k)$$
$$= -C_R(\theta_i - \theta_{i-1}) + s(\theta_{i-1}) - s(\theta_i)$$
$$+ [\Delta(\theta_i - \theta_{i-1}) - (s(\theta_{i-1}) - s(\theta_i))]$$
$$\cdot \{[\mathsf{E}(D - n_{i,k} - N_{1,i-1})^+ - \mathsf{E}(D - n_{i,k} - N_{1,i-1} - 1)^+]\}, \quad (9.4)$$

which is decreasing in $n_{i,k}$, and in particular, decreasing in n_j for all $j \ge i$; furthermore,

$$\Pi(n_1, \cdots, n_{i-1}, n_i, \cdots, n_k) - \Pi(n_1, \cdots, n_{i-1}, n_i + 1, \cdots, n_k)$$
$$= -(C_I + C_R \theta_i)$$
$$+ \sum_{j=1}^{i} \left\{ [s(\theta_{j-1}) - s(\theta_j)] + [\Delta(\theta_j - \theta_{j-1}) - (s(\theta_{j-1}) - s(\theta_j))] \right.$$
$$\left. \cdot [\mathsf{E}(D - n_{j,k} - N_{1,j-1})^+ - \mathsf{E}(D - n_{j,k} - N_{1,j-1} - 1)^+] \right\}, \quad (9.5)$$

which is decreasing in $n_{j,k}$, for all $j = 1, ..., k$, and hence decreasing in n_j, for all j.

Proof. Making use of the relation

$$\min\{a, b\} = a - (a - b)^+ = a - (b - a)^-$$

and noticing that $(a^+ - b)^+ = (a - b)^+$ for $b \ge 0$, we rewrite the terms on the right-hand side of (9.2) as follows:

$$
\min\{n_{1,k}, D\}C_g
$$
$$
+ \sum_{i=1}^{k} \min\{N_i - n_i, (D - n_{i,k} - N_{1,i-1})^+\}(C_g + \theta_i \Delta)
$$
$$
= \quad [D - (D - n_{1,k})^+]\, C_g
$$
$$
+ \sum_{i=1}^{k} \{(D - n_{i,k} - N_{1,i-1})^+
$$
$$
- [(D - n_{i,k} - N_{1,i-1})^+ - (N_i - n_i)]^+\}(C_g + \theta_i \Delta)
$$
$$
= \quad [D - (D - n_{1,k})^+]C_g
$$
$$
+ \sum_{i=1}^{k} [(D - n_{i,k} - N_{1,i-1})^+ - (D - n_{i+1,k} - N_{1,i})^+](C_g + \theta_i \Delta)
$$
$$
= \quad [D - (D - N_{1,k})^+]C_g
$$
$$
+ \sum_{i=1}^{k} (D - n_{i,k} - N_{1,i-1})^+ (\theta_i - \theta_{i-1})\Delta - (D - N_{1,k})^+ \theta_k \Delta
$$

and

$$
- \sum_{i=1}^{k} [N_i - n_i - (D - n_{i,k} - N_{1,i-1})^+]^+ s(\theta_i)
$$
$$
= \quad \sum_{i=1}^{k} \{\min[N_i - n_i, (D - n_{i,k} - N_{1,i-1})^+] - (N_i - n_i)\} s(\theta_i)
$$
$$
= \quad \sum_{i=1}^{k} \{(D - n_{i,k} - N_{1,i-1})^+ - [(D - n_{i,k} - N_{1,i-1})^+
$$
$$
- (N_i - n_i)]^+ - (N_i - n_i)\} s(\theta_i)
$$
$$
= \quad \sum_{i=1}^{k} \{(D - n_{i,k} - N_{1,i-1})^- - (D - n_{i+1,k} - N_{1,i})^-\} s(\theta_i)
$$
$$
= \quad \sum_{i=1}^{k} (D - n_{i,k} - N_{1,i-1})^- [s(\theta_i) - s(\theta_{i-1})]
$$
$$
+ (D - n_{1,k})^- s(\theta_0) - (D - N_{1,k})^- s(\theta_k).
$$

Hence, (9.2) becomes:

$$
W(D, n_1, \cdots, n_k)
$$
$$
= \quad [D - (D - N_{1,k})^+]C_g - (D - N_{1,k})^+ \theta_k \Delta
$$

$$+ \sum_{i=1}^{k} (D - n_{i,k} - N_{1,i-1})^+ (\theta_i - \theta_{i-1}) \Delta$$

$$- \sum_{i=1}^{k} (D - n_{i,k} - N_{1,i-1})^- [s(\theta_{i-1}) - s(\theta_i)] - (D - N_{1,k})^- s(\theta_k)$$

$$= C_g D - (C_g + \Delta \theta_k + s(\theta_k))(D - N_{1,k})^+ + (D - N_{1,k})s(\theta_k)$$

$$+ \sum_{i=1}^{k} (D - n_{i,k} - N_{1,i-1})^+ [(\theta_i - \theta_{i-1})\Delta - (s(\theta_{i-1}) - s(\theta_i))]$$

$$+ \sum_{i=1}^{k} (D - n_{i,k} - N_{1,i-1})(s(\theta_{i-1}) - s(\theta_i)). \qquad (9.6)$$

(Recall $\theta_0 = 0$.) Making use of (9.6), we can derive

$$W(D, n_1, \cdots, n_{i-1} + 1, n_i, \cdots, n_k)$$
$$-W(D, n_1, \cdots, n_{i-1}, n_i + 1, \cdots, n_k)$$
$$= s(\theta_{i-1}) - s(\theta_i) + [\Delta(\theta_i - \theta_{i-1}) - (s(\theta_{i-1}) - s(\theta_i))]$$
$$\cdot [(D - n_{i,k} - N_{1,i-1})^+ - (D - n_{i,k} - N_{1,i-1} - 1)^+], \quad (9.7)$$

noticing that the two Ws only differ at the ith term (in the summation). By the condition in (9.1) and the fact that $(D - x)^+$ is a convex function of x, (9.7) is decreasing in $n_{i,k}$.

The proof is then completed by incorporating the results, (9.6) and (9.7) in particular, into (9.3). $\square$

Remark 9.2 It is worthwhile to point out several facts that relate to the properties summarized in Proposition 9.5. The property in (9.5) clearly implies the convexity of $\Pi(n_1, \cdots, n_k)$ in each component. To appreciate the property in (9.4), note that the difference is independent of the value of n_{i-1}, in particular we have

$$\Pi(n_1, \cdots, n_{i-1} + 1, n_i, \cdots, n_k) - \Pi(n_1, \cdots, n_{i-1}, n_i + 1, \cdots, n_k)$$
$$= \Pi(n_1, \cdots, n_{i-1} + 2, n_i, \cdots, n_k)$$
$$-\Pi(n_1, \cdots, n_{i-1} + 1, n_i + 1, \cdots, n_k),$$

which can be rewritten as follows:

$$[\Pi(n_1, \cdots, n_{i-1}, n_i, \cdots, n_k) - \Pi(n_1, \cdots, n_{i-1}, n_i + 1, \cdots, n_k)]$$
$$-[\Pi(n_1, \cdots, n_{i-1}, n_i, \cdots, n_k) - \Pi(n_1, \cdots, n_{i-1} + 1, n_i, \cdots, n_k)]$$
$$= [\Pi(n_1, \cdots, n_{i-1} + 1, n_i, \cdots, n_k)$$
$$-\Pi(n_1, \cdots, n_{i-1} + 1, n_i + 1, \cdots, n_k)]$$
$$-[\Pi(n_1, \cdots, n_{i-1} + 1, n_i, \cdots, n_k) - \Pi(n_1, \cdots, n_{i-1} + 2, n_i, \cdots, n_k)].$$

The difference in the second bracket on the left-hand side dominates its counterpart on the right-hand side, due to convexity. Therefore, we must have

$$\Pi(n_1, \cdots, n_{i-1}, n_i, \cdots, n_k) - \Pi(n_1, \cdots, n_{i-1}, n_i + 1, \cdots, n_k)$$
$$\geq \quad \Pi(n_1, \cdots, n_{i-1} + 1, n_i, \cdots, n_k) - \Pi(n_1, \cdots, n_{i-1} + 1, n_i + 1, \cdots, n_k),$$

which is nothing but *supermodularity* (refer to §2.3). In other words, the properties of Π as revealed in Proposition 9.1 imply supermodularity and componentwise convexity. It can be verified, however, that these properties in general will *not* guarantee the optimality of the greedy algorithm given later. Of course, these properties are weaker than (implied by) those in Proposition 9.1. Furthermore, the particular form of the difference expressions in (9.4 and 9.5) also plays an important role, as will become evident.

From (9.4), we can write

$$g_i(n_{i,k})$$
$$:= \quad \Pi(n_1, \cdots, n_{i-1} + 1, n_i, \cdots, n_k) - \Pi(n_1, \cdots, n_{i-1}, n_i + 1, \cdots, n_k)$$

and define

$$K_i := \min \{ n_{i,k} \leq N_{i,k} : g_i(n_{i,k}) < 0 \}.$$

By regulation, define $K_i := N_{i,k}$ if no $n_{i,k} \leq N_{i,k}$ satisfies $g_i(n_{i,k}) < 0$. Clearly, K_i is the smallest value for $n_{i,k}$ so that it becomes more desirable to inspect one more unit from batch $i - 1$ than to inspect one more unit from batch i in any state $(n_1, n_2, \cdots, n_{i-1}, n_i, \cdots, n_k)$.

Lemma 9.3 Suppose $n_{i-1} < N_{i-1}$ and $n_i < N_i$. Then it is more desirable in state $(n_1, \cdots, n_{i-1}, n_i, \cdots, n_k)$ to inspect one more unit from batch $i-1$ than to inspect *any* more units from batch i if and only if $n_{i,k} \geq K_i$.

Proof. From the definition of K_i, we know in state $(n_1, \cdots, n_{i-1}, n_i, \cdots, n_k)$, that it is more desirable to inspect one more unit from batch $i - 1$ than to inspect one more unit from batch i if and only if $n_{i,k} \geq K_i$. But this also holds if we inspect more than one unit from batch i, because this will only increase the value of $n_{i,k}$. $\square$

Corollary 9.4 Suppose $n_\ell = N_\ell$, for $\ell = j, j+1, \cdots, i-1$, $1 \leq j < i \leq k$. That is, all the units in batch j through batch $i - 1$ have been inspected. And $n_{j-1} < N_{j-1}$, $n_i < N_i$. Then Lemma 9.3 holds with batch $j - 1$ replacing batch $i - 1$. That is, it is more desirable to inspect one more unit from batch $j - 1$ than to inspect *any* more units from batch i if and only if $n_{i,k} \geq K_i$.

Proof. We can merge batches j through $i - 1$ into a single nondefective batch and remove them from further consideration. This will not affect the values of $n_{i,k}$ or K_i. The desired conclusion then follows from Lemma 9.3. $\square$

Remark 9.5 Two points are worth mentioning here:

(i) From (9.5), we know that as more units are inspected in batch i, the cost reduction diminishes. Furthermore, as more units are inspected in batch i, $n_{i,k}$ increases, which, in turn, increases the desirability of switching to inspecting batch $i - 1$, following Lemma 9.3. Once $n_{i,k}$ reaches K_i, switching to $i - 1$ becomes more desirable, and this remains so even when more units from batch $i - 1$ are inspected, because $n_{i,k}$ is independent of n_{i-1}.

(ii) On the other hand, switching inspection to batch $i - 1$ might never be desirable. For example, this can happen when the cost reduction in (9.5) becomes negative but we still have $n_{i,k} < K_i$. In this case, it becomes more desirable to simply stop inspection rather than inspecting any more units from either batch i or batch $i - 1$.

9.3   Optimal Solution to the Inspection Problem

We will assume that the salvage value $s(\theta)$ is convex in θ and decreasing in θ as assumed earlier. Consequently, $\frac{s(\theta_{i-1}) - s(\theta_i)}{\theta_i - \theta_{i-1}}$ is decreasing in i.

From the definition of $g_i(n_{i,k})$, making use of (9.4), we know that

$$g_i(n_{i,k}) < 0$$

is equivalent to

$$C_R - \frac{s(\theta_{i-1}) - s(\theta_i)}{\theta_i - \theta_{i-1}}$$
$$> \quad [\Delta - \frac{s(\theta_{i-1}) - s(\theta_i)}{\theta_i - \theta_{i-1}}] E[(D - y)^+ - (D - y - 1)^+] \qquad (9.8)$$

with $y := n_{i,k} + N_{1,i-1}$. From (9.1), we know the first factor on the right-hand side is nonnegative. To start, suppose it is positive. (Later it will become evident that the same argument applies to the case when this factor is zero.) Then the inequality in (9.8) is equivalent to the following:

$$\frac{C_R - \frac{s(\theta_{i-1}) - s(\theta_i)}{\theta_i - \theta_{i-1}}}{\Delta - \frac{s(\theta_{i-1}) - s(\theta_i)}{\theta_i - \theta_{i-1}}} > E[(D - y)^+ - (D - y - 1)^+]. \qquad (9.9)$$

A direct verification shows that when $s(\theta)$ is a decreasing and convex function, the left-hand side of (9.9) is increasing in i, taking into account that $\frac{s(\theta_{i-1}) - s(\theta_i)}{\theta_i - \theta_{i-1}}$ is decreasing in i and $C_R \leq \Delta$. On the other hand, as i increases, the right-hand side of (9.9) decreases (again, due to the convexity of $(D - x)^+$). In fact, because the right-hand side decreases as y increases, its decrease can be carried out in a more detailed manner: say, from the state $(n_1, ..., n_i, ..., n_k)$, we can first increase n_i to N_i, and then n_{i+1} to N_{i+1}, and so forth.

Therefore, starting from the zero state, $(0, ..., 0)$, we increase each component n_i, following the order $i = 1, ..., k$, from 0 to N_i. Let i^* be the first i index and $\hat{n}_{i^*}$ be the smallest corresponding component value such that $y^* = \hat{n}_{i^*} + N_{1,i^*-1}$ is the smallest y that satisfies (9.9). For the time being, suppose such a y^* does exist. This has the following implications:

(a) After $\hat{n}_{i^*}$ units from batch i^* are inspected, it becomes more desirable to switch to inspecting units from batch $i^* - 1$. In other words, $K_{i^*} = y^* - N_{1,i^*-1} = \hat{n}_{i^*}$.

(b) We can then inspect each batch $i = i^* - 1, ..., 1$ in decreasing order of i; and there is no need to switch to batch $i - 1$ until all units of batch i have been inspected. This is because in this range $(i < i^*)$ the left-hand side of (9.9) is dominated by the right-hand side (i.e., the inequality is satisfied in the reverse direction), which implies, via (9.4 and 9.5), that the cost reduction in (9.5) dominates the same cost reduction when the index i is changed to $i - 1$. Hence, in this case, $K_i = y^* - N_{1,i-1}$, for $i = 1, ..., i^* - 1$.

(c) No unit should be inspected from any batch $i^* + 1, ..., k$, until all the batches $1, ..., i^*$ have been inspected, because for $i > i^*$, (9.9) is always satisfied. In other words, $K_i = 0$, for $i = i^* + 1, ..., k$.

(d) In view of Remark 9.5 (ii), however, in cases (a) and (b), as we increase the number of inspected units in each batch, we still need to make sure that the cost reduction in (9.5) is positive. Once the cost reduction becomes nonpositive, inspection should be terminated for all batches.

(e) On the other hand, if the cost reduction in (9.5) stays positive and all units in the batches $i = 1, ..., i^* - 1$ have been inspected, then we need to return to batch i^*. Note that switching to $i^* - 1$ in (a) was due to the positive cost reduction in (9.4). Now, although this cost reduction remains positive, switching to a lower indexed batch becomes out of the question, because all units in those batches have been inspected. Hence, inspecting more units from batch i^* is warranted, as long as the cost reduction in (9.5) is positive. In fact, we need to consider the batches $i > i^*$ as well, in increasing order of i, for the same reason as in (b), because here the cost reduction in (9.5) dominates the same cost reduction when the index i is changed to $i + 1$.

There are cases in which the inequality in (9.8) just cannot be satisfied. For example, if the left-hand side of (9.8) is negative, then the inequality cannot hold, because both factors on the right-hand side are nonnegative.

Another case is when y^* (as defined earlier) does not exist: even when i is increased to k and y is increased to its upper limit $N_{1,k}$, the left-hand side of (9.9) is dominated by the right-hand side. Both of these instances imply that inspecting batch i is always more preferable than inspecting batch $i-1$ for any i or $K_i = N_{i,k}$; and hence, the optimal solution is to inspect the batches in decreasing order of i and to stop inspection whenever the cost reduction in (9.5) becomes nonpositive. A third case is when the salvage value $s(\theta)$ is a linear function, and

$$\Delta = \frac{s(\theta_{i-1}) - s(\theta_i)}{\theta_i - \theta_{i-1}} \leq C_R \tag{9.10}$$

for all i. This, along with the assumption that $C_R \leq \Delta$, implies that the inequality in (9.10) holds as an equality, and hence both sides of (9.8) are zero. In fact, in this case the right-hand side of (9.5) becomes $-C_I - C_R\theta_0 \leq 0$. This leads to the optimality of the trivial "do-nothing" solution, $(0, ..., 0)$. Correspondingly, $K_i = 0$ for all $i = 1, ..., k$.

We now return to the case when the first factor on the right-hand side of (9.8) is zero for some i. Because this factor is nonnegative and increasing in i, let $\bar{i}$ be the largest i for which it stays at zero. Then, the left-hand side of (9.8) is nonpositive for all $i \leq \bar{i}$, because $C_R \leq \Delta$. That is, (9.8) is not satisfied for all $i \leq \bar{i}$. For $i > \bar{i}$, on the other hand, the factor in question becomes positive and we can repeat the earlier argument based on (9.9). In particular, we know $i^* > \bar{i}$ in this case.

To summarize, we have the following.

Theorem 9.6 Suppose the salvage value $s(\theta)$ is a convex and decreasing function. Let $i^* \leq k$ be the smallest batch index and let $\hat{n}_{i^*} \leq N_{i^*}$ be the smallest corresponding component value, such that $y^* = \hat{n}_{i^*} + N_{1,i^*-1}$ is the smallest y that satisfies (9.9).

(i) When such a y^* exists, the optimal solution is either

$$(0, ..., 0, n'_h, N_{h+1}, ..., N_{i^*-1}, n'_{i^*}, 0, ..., 0)$$

or

$$(N_1, ..., N_{j^*-1}, n'_{j^*}, 0, ..., 0),$$

where

- n'_{i^*} is equal to either $\hat{n}_{i^*}$ or the smallest n_i (with $i = i^*$) value that results in a nonpositive cost reduction in (9.5), if this value is less than $\hat{n}_{i^*}$;
- $1 \leq h \leq i^*$, $n'_h \leq N_h$, and n'_h is the smallest n_i (with $i = i^*$) value that results in a nonpositive cost reduction in (9.5);
- $j^* \geq i^*$ and $n'_{j^*} \leq N_{j^*}$ is the smallest n_i (with $i = i^*$) value that results in a nonpositive cost reduction in (9.5), or, if no such value exists, then $n'_{j^*} = N_k$ with $j^* = k$.

(ii) If y^* as defined does not exist or if the left-hand side of (9.8) is negative for all i, then the optimal solution is obtained as follows: inspect the batches in decreasing order of i starting from $i = k$ and stop inspection whenever the cost reduction in (9.5) becomes nonpositive. In this case, the optimal solution is

$$(0, ..., 0, n'_{j^*}, N_{j^*+1}, ..., N_k)$$

with $1 \leq j^* \leq k$.

(iii) If the cost data satisfy the relation in (9.10), then the optimal solution is $(0, ..., 0)$, i.e., inspect no units.

Proof. (i) It is clear from the construction of both solutions that a decrease of any of the positive components will result in a sacrifice of some positive cost reduction. This includes reducing some positive component while increasing another component (by the same amount)—a procedure that we shall refer to as 'shifting'. It suffices to argue that none of the components can be increased either; we only need to examine those components that have not reached the given batch sizes.

Consider the first solution. Clearly, we cannot increase n'_{i^*} without resulting in a cost increase (via (9.4 and 9.5)). Suppose $n_j > 0$ for $j = i^* + 1$. Then, clearly, $y = n_{j,k} + N_{1,j-1} > y^*$ satisfies (9.9) (with $i = j$). Therefore, the overall cost will decrease, following Lemma 9.3 and Corollary 9.4, if we shift one unit from batch j to batch i^* (if $n'_{i^*} < N_{i^*}$) or to batch $i \leq h$ (if $n'_{i^*} = N_{i^*}$). A similar argument applies if $n_j > 0$ for $j > i^* + 1$ through repeatedly shifting units from batch j to batches with lower indices.

In the second solution, increasing any component $j \geq j^*$ will further decrease the right-hand side of (9.5), resulting in a nonpositive cost reduction—beyond what results in the case of $\hat{n}_{j^*}$.

(ii) In this case, as in (i), decreasing any positive component of the optimal solution, including shifting it to some other component, will result in a cost increase. On the other hand, because the optimal solution is reached when the cost reduction in (9.5) ceases to be positive, increasing any component of the optimal solution will result in a nonpositive cost reduction, through increasing the $n_{j,k} + N_{1,j-1}$ value in (9.5).

(iii) In this case, any solution that has a positive component cannot be optimal, because reducing the positive component by one unit will result in a cost reduction of $C_I + C_R \theta_0$ via the discussion following (9.10). $\square$

Remark 9.7 The K_i values in the three cases of Theorem 9.6 are:

(i) $K_i = y^* - N_{1,i-1}$, for $i = 1, ..., i^*$ (in particular, $K_{i^*} = y^* - N_{1,i^*-1} = \hat{n}_{i^*}$); and $K_j = 0$, for $j = i^* + 1, ..., k$.

(ii) $K_i = N_{i,k}$, for $i = 1, ..., k$.

	Data and Optimal Solutions									
Type i	1	2	3	4	5	6	7	8	9	10
Batch size N_i	20	20	20	20	20	20	20	20	20	20
Defective rate θ_i	.03	.08	.12	.15	.17	.20	.22	.25	.30	.40
$(n_1^*, \cdots, n_{10}^*)$ (N)	0	0	17	20	20	20	20	20	0	0
$(n_1^*, \cdots, n_{10}^*)$ (U)	0	0	12	20	20	20	20	20	3	0
$(n_1^*, \cdots, n_{10}^*)$ (P)	0	4	20	20	20	20	20	17	0	0

TABLE 9.1. Optimal inspection policy.

	Objective Value		
	Optimal Inspection	Full Inspection	Zero Inspection
Normal	248.5	276.0	282.7
Uniform	245.1	273.1	276.5
Poisson	244.2	275.4	280.4

TABLE 9.2. Objective values under different inspection policies.

(iii) $K_i = 0$, for $i = 1, ..., k$.

Note that in all three cases, K_i is decreasing in i. As evident from the preceding discussions, this turns out to be the key to the threshold structure of the optimal solution in Theorem 9.6.

Example 9.8 Consider the following inspection problem. Suppose we have a total of ten types of products with batch sizes and defective rates listed in Table 9.1. The cost data are:

$$C_d = 7.0, \quad C_g = 1.0, \quad C_I = 0.2, \quad C_R = 2.9;$$

and the salvage value is a convex function, $s(\theta) = 1.5 - 2.5\theta + 3\theta^2$. Consider three types of demand distributions:

(a) normal (N) with mean 150 and standard deviation 20,

(b) uniform (U) over the interval $(115.36, 184.64)$,

(c) Poisson (P) with mean 150.

Note that all three distributions have the same mean; in addition, the normal and uniform distributions have the same standard deviation. The optimal solutions, $(n_1^*, \cdots, n_{10}^*)$, under the three demand distributions are listed in Table 9.1. The corresponding optimal objective values are listed in Table 9.2, in comparison with those under full (100%) inspection and zero inspection.

9.4   Optimal Replenishment Quantities

Here we extend the earlier model to include the batch sizes, N_i, $i = 1, ..., k$, as decision variables. Specifically, we want to decide the order quantity of each product type, taking into account that these products will be inspected, following the optimal rule discussed in the earlier sections and then used to supply demand.

In addition to the cost data in §9.1, there is a penalty cost, C_P, for each unit of shortage (unfilled demand). Assume

$$C_P - s(\theta_k) \geq C_g + \Delta\theta_k, \qquad (9.11)$$

which implies

$$C_P - C_g \geq \Delta\theta_k + s(\theta_k) \geq \Delta\theta_i + s(\theta_i)$$

for all $i \leq k$. This simply guarantees that any demand will be supplied if there is a product available. There is also a purchasing cost, $c(\theta_i)$, for each unit of product i, $i = 1, ..., k$. We assume that the purchasing cost net the salvage value, $c(\theta) - s(\theta)$, is a decreasing and convex function of the defective rate θ. Note that this implies the decreasing convexity of $c(\theta)$, because $s(\theta)$ is a decreasing and convex function, as assumed earlier.

Furthermore, we assume that type-1 products are of perfect quality: $\theta_1 = 0$ (in contrast with $\theta_0 = 0$ in the previous sections). This way, we can address the tradeoff between purchasing perfect units at a higher cost and purchasing lower-quality units but spending more on inspection and repair. Note that ordering products with defective rates $\theta = 0$ and $\theta > 0$ corresponds, respectively, to the "selective purchase" and the "blind purchase" in [53].

Let $v(N_1, \cdots, N_k)$ denote the optimal cost function considered in the last section, given the batch sizes, N_i units of product i, for $i = 1, 2, \cdots, k$. Let $V(N_1, \cdots, N_k)$ denote the new objective (cost) function here, i.e.,

$$V(N_1, \cdots, N_k)$$

$$:= \sum_{i=1}^{k} N_i c(\theta_i) + v(N_1, \cdots, N_k) + C_P \mathsf{E}(D - N_{1,k})^+. \qquad (9.12)$$

We want to find $(N_1, \cdots, N_k)$ to minimize the V function. We need the following lemma.

Lemma 9.9 Let

$$\ell^* = arg \min_{2 \leq i \leq k} \{i : c(\theta_i) + C_R\theta_i\}. \qquad (9.13)$$

Suppose $c(0) \leq c(\theta_{l^*}) + C_I + C_R\theta_{\ell^*}$. Then it is optimal not to inspect any unit. Hence, the cost function in (9.12) is reduced to

$$V(N_1, \cdots, N_k)$$

$$= \sum_{i=1}^{k} N_i c(\theta_i) + \Pi(0, \cdots, 0; N_1, \cdots, N_k)$$
$$+ C_P \mathsf{E}(D - N_{1,k})^+, \tag{9.14}$$

where $\Pi(0, \cdots, 0; \cdot) = \Pi(0, \cdots, 0)$ in (9.3). Furthermore,

$$V(N_1, \cdots, N_i, \cdots, N_k) - V(N_1, \cdots, N_i + 1, \cdots, N_k)$$
$$= -c(\theta_i) + s(\theta_i) + [C_P - C_g - \Delta\theta_k - s(\theta_k)]$$
$$\cdot[\mathsf{E}(D - N_{1,k})^+ - \mathsf{E}(D - N_{1,k} - 1)^+]$$
$$+ \sum_{j=i+1}^{k} [\Delta(\theta_j - \theta_{j-1}) - (s(\theta_{j-1}) - s(\theta_j))]$$
$$\cdot[\mathsf{E}(D - N_{1,j-1})^+ - \mathsf{E}(D - N_{1,j-1} - 1)^+], \tag{9.15}$$

which is decreasing in N_j, for any $j = 1, ..., k$; and

$$V(N_1, \cdots, N_{i-1} + 1, N_i, \cdots, N_k) -$$
$$V(N_1, \cdots, N_{i-1}, N_i + 1, \cdots, N_k)$$
$$= c(\theta_{i-1}) - c(\theta_i) - [s(\theta_{i-1}) - s(\theta_i)]$$
$$-[\Delta(\theta_i - \theta_{i-1}) - (s(\theta_{i-1}) - s(\theta_i))]$$
$$\cdot[\mathsf{E}(D - N_{1,i-1})^+ - \mathsf{E}(D - N_{1,i-1} - 1)^+], \tag{9.16}$$

which is increasing in $N_{1,i-1}$.

Proof. First, it is easy to see that ℓ^* is the only product type (among types 2 through k) we may consider for inspection. (Type 1 does not need inspection anyway.) This is because, instead of ordering N_i and N_{ℓ^*} units for any $i \neq 1, \ell^*$, and inspecting n_i and n_{ℓ^*} units, respectively, we can order $N_i - n_i$ and $N_{\ell^*} + n_i$ units, and then inspect 0 and $n_{\ell^*} + n_i$ units for type i and type ℓ^*, respectively. This way, the overall cost reduction (in V) is

$$n_i \{[c(\theta_i) - c(\theta_{\ell^*})] + C_R(\theta_i - \theta_{l^*})\},$$

which is nonnegative, following the definition of ℓ^* in (9.13).

Hence, we can obtain a perfect unit, at an expected cost of $c(\theta_{\ell^*}) + C_I + C_R\theta_{\ell^*}$, by ordering and inspecting one unit of type ℓ^*. Alternatively, we can get a perfect unit, at a cost $c(0)$, by ordering one unit of product 1, because $\theta_1 = 0$. This alternative is certainly preferred when

$$c(0) \leq c(\theta_{l^*}) + C_I + C_R\theta_{l^*}.$$

In this case we will never inspect any unit. Should we choose to inspect any unit in types 2 through k, we would prefer not to order the unit, but instead replace it by a type ℓ^* unit and then inspect the latter. But then we would further prefer to replace the type ℓ^* unit by a type 1 unit and forgo inspection.

To establish the properties for (9.15) and (9.16), letting

$$(n_1, ..., n_k) = (0, ..., 0)$$

in (9.3) and (9.6), we have

$$
\begin{aligned}
&\Pi(0, ..., 0; N_1, ..., N_k) \\
=\ &C_g \mathsf{E}[D - (D - N_{1,k})^+] - \Delta\theta_k \mathsf{E}(D - N_{1,k})^+ - s(\theta_k)\mathsf{E}(D - N_{1,k})^- \\
&+ \sum_{j=1}^{k} \{\Delta(\theta_j - \theta_{j-1})\mathsf{E}(D - N_{1,j-1})^+ \\
&\qquad - (s(\theta_{j-1}) - s(\theta_j))\mathsf{E}(D - N_{1,j-1})^-\}.
\end{aligned}
\tag{9.17}
$$

Substituting this into (9.14), we have

$$
\begin{aligned}
&V(N_1, \cdots, N_i, \cdots, N_k) - V(N_1, \cdots, N_i + 1, \cdots, N_k) \\
=\ &-c(\theta_i) + (C_P - C_g)[\mathsf{E}(D - N_{1,k})^+ - \mathsf{E}(D - N_{1,k} - 1)^+] \\
&+ \sum_{j=i+1}^{k} \{\ \Delta(\theta_j - \theta_{j-1})[\mathsf{E}(D - N_{1,j-1})^+ - \mathsf{E}(D - N_{1,j-1} - 1)^+] \\
&\qquad - (s(\theta_{j-1}) - s(\theta_j))[\mathsf{E}(D - N_{1,j-1})^- - \mathsf{E}(D - N_{1,j-1} - 1)^-]\ \} \\
&- \Delta\theta_k[\mathsf{E}(D - N_{1,k})^+ - \mathsf{E}(D - N_{1,k} - 1)^+] \\
&- s(\theta_k)[\mathsf{E}(D - N_{1,k})^- - \mathsf{E}(D - N_{1,k} - 1)^-].
\end{aligned}
$$

Collecting terms and making use of the identity $x^+ - x^- = x$, we can simplify the preceding to (9.15). The decreasing property follows from (9.1), (9.11) and the convexity of $(D - x)^+$. Next, because

$$
\begin{aligned}
&V(N_1, \cdots, N_{i-1} + 1, N_i, \cdots, N_k) - V(N_1, \cdots, N_{i-1}, N_i + 1, \cdots, N_k) \\
=\ &[V(N_1, \cdots, N_i, \cdots, N_k) - V(N_1, \cdots, N_i + 1, \cdots, N_k)] \\
&- [V(N_1, \cdots, N_{i-1}, \cdots, N_k) - V(N_1, \cdots, N_{i-1} + 1, \cdots, N_k)],
\end{aligned}
$$

applying (9.15) twice yields (9.16). The increasing property follows from the convexity of $(D - x)^+$ and (9.1). $\square$

Based on (9.16), we can write

$$
\begin{aligned}
&G_i(N_{1,i}) \\
:=\ &V(N_1, \cdots, N_i + 1, N_{i+1}, \cdots, N_k) - V(N_1, \cdots, N_i, N_{i+1} + 1, \cdots, N_k).
\end{aligned}
$$

Define

$$M_i := \min\{N_{1,i} : G_i(N_{1,i}) > 0\} \tag{9.18}$$

for $i = 1, \cdots, k - 1$; and $M_0 := 0$. Note that $G(N_{1,i}) > 0$ is equivalent to

$$\frac{[c(\theta_i) - s(\theta_i)] - [c(\theta_{i+1}) - s(\theta_{i+1})]}{\theta_{i+1} - \theta_i} \tag{9.19}$$

$$> \quad [\Delta - \frac{s(\theta_i) - s(\theta_{i+1})}{\theta_{i+1} - \theta_i}][\mathsf{E}(D - N_{1,i})^+ - \mathsf{E}(D - N_{1,i} - 1)^+], \tag{9.20}$$

which means that it becomes more desirable to order one more unit of type $i + 1$ than to order one more unit of type i.

To start, consider $i = 1$. The right-hand side of (9.19) is decreasing in $N_{1,1} \equiv N_1$. (Note that the first factor on the right-hand side is nonnegative, following (9.1).) Hence, when N_1 is large enough to satisfy the inequality, i.e., $N_1 = M_1$ following (9.18), we should stop ordering any more units of type 1, and switch to ordering type-2 units. Next, consider $i = 2$. Because $c(\theta) - s(\theta)$ is decreasing and convex in θ as assumed earlier, the left-hand side of (9.19) is decreasing in i. And, because $s(\cdot)$ is decreasing and convex, the first factor on the right-hand side of (9.19) is increasing in i. Hence, the smallest $N_{1,2}$ that satisfies the inequality, i.e., M_2 as denoted in (9.18), will be no less than M_1; and the order size for type-2 units is up to $M_2 - M_1$: after that limit is reached we should switch to ordering type-3 units, and so forth. In general, when type i has been ordered to its maximum, $M_i - M_{i-1}$, we should switch to ordering type $i + 1$.

On the other hand, before $N_{1,i}$ reaches M_i, (9.19) holds in the reverse direction, which means the right-hand side of (9.16) is nonpositive. This, in turn, implies

$$[V(N_1, \cdots, N_{i-1}, N_i, \cdots, N_k) - V(N_1, \cdots, N_{i-1} + 1, N_i, \cdots, N_k)]$$
$$\geq \quad [V(N_1, \cdots, N_{i-1}, N_i, \cdots, N_k) - V(N_1, \cdots, N_{i-1}, N_i + 1, \cdots, N_k)].$$

That is, until $N_{1,i}$ reaches M_i or until type i has been ordered to its maximum, $M_i - M_{i-1}$, there is no need to switch to ordering type $i + 1$.

From (9.15), we know that as more units of product i are ordered, the cost reduction decreases. Hence, in ordering each additional unit of type i, even before reaching the limit $M_i - M_{i-1}$, we need to make sure that the cost reduction in (9.15) is positive. Should this cost reduction become nonpositive before the limit $M_i - M_{i-1}$ is reached, we should stop ordering altogether—not just type i, but all types $j > i$. Hence, (9.15) plays the same role as (9.5) in the earlier model. If $N_i = M_i - M_{i-1}$ for $i = 1, 2, \cdots, k - 1$, then N_k is determined by (9.15) with $i = k$, i.e., it is the smallest order quantity of product k so that (9.15) becomes nonpositive.

Finally, a special case is of particular interest: when both $c(\theta)$ and $s(\theta)$ are linear functions, it is easy to see from (9.19) that $M_1 = M_2 = \cdots = M_{k-1}$ in this case. Consequently, product $2, 3, \cdots, k - 1$ will not be ordered, and N_1 and (possibly) N_k are the only nonzero components in the optimal solution. In other words, it is optimal to use only two supply sources, 1 and k. To summarize, we have the following.

Theorem 9.10 Suppose $c(0) \leq c(\theta_{\ell^*}) + C_I + C_R \theta_{l^*}$. Then the optimal solution to the order quantities is obtained as follows:

- order the units in increasing order of the type index i, starting from $i = 1$;

- every time a unit of type i is ordered, check whether the cost reduction in (9.15) stays positive; if not, stop ordering any more units from any type;

- as long as the cost reduction in (9.15) stays positive, keep ordering type-i units, until $N_i = M_i - M_{i-1}$, then switch to ordering type $i + 1$ units, for $i = 1, 2, \cdots, k - 1$. Here M_i follows the specification in (9.18);

- if $N_i = M_i - M_{i-1}$ for $i = 1, 2, \cdots, k - 1$, then keep ordering type-k units until (9.15), with $i = k$, becomes nonpositive.

Furthermore, following Lemma 9.9, in this case it is optimal not to inspect any unit from any type. When $c(\theta)$ and $s(\theta)$ are linear functions, $N_i = 0$ for $i = 2, ..., k - 1$. That is, it is optimal to use only two supply sources, 1 and k.

Proof. Following the specification in the theorem, we can write the optimal solution as $(N_1^*, ..., N_{i^*-1}^*, N_{i^*}^*, 0, ..., 0)$, where $N_i^* \leq M_i - M_{i-1}$ for $i \leq i^*$ and $1 \leq i^* \leq k$ is the smallest i value in (9.15) that makes its right-hand side nonpositive. (Recall, from the preceding analysis, this right-hand side is decreasing in i.) Similar to the argument in the proof of Theorem 9.6, it is clear that decreasing any of the positive component amounts will forgo some positive cost reduction. On the other hand, increasing the value of any component $i \leq i^* - 1$ is not as good as increasing that of $i + 1$, because the right-hand side of (9.16) is positive; and increasing the value of any component $i \geq i^*$ will result in a cost increase via (9.15). $\square$

Next, consider the case of $c(0) > c(\theta_{\ell^*}) + C_I + C_R \theta_{\ell^*}$. In this case, it does not pay to order any unit of product 1. Instead we are better off ordering units of type ℓ^* and converting them into nondefective units via inspection and possible repair. Denote the units so obtained as of type $1'$. Then we can solve the problem following Theorem 9.10, treating product $1'$ as product 1, with a zero defect rate, and with the purchasing cost $c(\theta_1) = c(0)$ replaced by $c(\theta_{\ell^*}) + C_I + C_R \theta_{\ell^*}$.

Theorem 9.11 Suppose $c(0) > c(\theta_{l^*}) + C_I + C_R \theta_{l^*}$. Then order zero units of type 1. Instead, replace type 1 by a type $1'$, which has zero defective rate, and a unit purchasing cost of $c(\theta_{\ell^*}) + C_I + C_R \theta_{\ell^*}$, with ℓ^* following (9.13). Follow Theorem 9.10 to derive the order quantities. Type $1'$ units

Type i	Data and Optimal Solutions											Obj.Val.
	1′	2	3	4	5	6	7	8	9	10	11	
Defective rate θ_i	.00	.03	.08	.12	.15	.17	.20	.22	.25	.30	.40	
$(N_1^*, N_2^*, \cdots, N_{11}^*)$ (N)	133	14	5	4	2	3	2	3	4	0	0	643.5
$(N_1^*, N_2^*, \cdots, N_{11}^*)$ (U)	128	17	7	5	3	3	3	3	4	0	0	638.3
$(N_1^*, N_2^*, \cdots, N_{11}^*)$ (P)	139	9	3	2	2	1	2	1	3	0	0	626.9

TABLE 9.3. Optimal order quantities.

are then ordered from type ℓ^*, with all units inspected (and repaired if necessary). Inspect no unit from any other types. When $c(\theta)$ and $s(\theta)$ are linear functions, it is optimal to only order from two supply sources, 1′ and k.

Remark 9.12 The dual sourcing result in Theorems 9.10 and 9.11 indicates that ordering from a single source is, in general, suboptimal, even with linear cost and linear salvage value. The intuitive reason is this: as we assume $c(\theta_i) - s(\theta_i)$ to be decreasing in i, units with a better quality (naturally) also have a higher net cost (i.e., cost minus salvage value). Hence, it does not pay to order a better-quality product that can not be used to supply demand. Because demand is random, there is always a possibility that some units ordered will be left over as surplus inventory. Hence, it is more desirable, following the optimality of dual sourcing, to order some units for possible backup from a second low-quality/low-cost source.

Finally, suppose there is *no* perfect type like product 1 to start. This clearly corresponds to the case in Theorem 9.11. That is, we can always pay a unit purchasing cost of $c(\theta_{\ell^*}) + C_I + C_R\theta_{\ell^*}$ to obtain a perfect unit, by ordering from type ℓ^* along with inspection and possible repair. This way, we have effectively created a perfect type. Another way to view this case is to set $c(0) = \infty$, signifying the unavailability of a perfect type. Then Theorem 9.11 naturally applies.

Example 9.13 Consider the problem in Example 9.8. In addition to the data given there, we have the penalty cost, $C_P = 10.0$, and the ordering cost, $c(\theta) = 3.0 - 5\theta + 6\theta^2$ for $\theta > 0$. There is also a perfect type, indexed as $i = 1$, with ordering cost $c(0) = 3.1$, and salvage value $s(0) = 3.1/2$. Accordingly, here we reindex the original ten product types as $2, 3, \cdots, 11$. We can identify $\ell^* = 6$ and $c(0) > c(\theta_{l^*}) + C_I + C_R\theta_{l^*}$. Hence, Theorem 9.11 applies, and no perfect unit should be ordered. For the three types of demand distributions in Example 9.8, the results are summarized in Table 9.3.

Note that here type 1′ units are obtained by ordering type ℓ^*. For example, when demand follows the normal distribution, we should order 136 units of type 6; of them 133 units are inspected and repaired (if necessary).

Next, suppose there are only two types available in Example 9.13, 2 and 11, the best and worst types. Then the optimal order quantities and

	Data and Optimal Solutions			Obj.Val.
Type i	$1'$	2	11	
Defective rate θ_i	.00	.03	.40	
$(N_{1'}^*, N_2^*, N_{11}^*)$ (N)	0	164	5	647.2
$(N_{1'}^*, N_2^*, N_{11}^*)$ (U)	0	167	5	641.9
$(N_{1'}^*, N_2^*, N_{11}^*)$ (P)	0	158	3	630.2

TABLE 9.4. Two supply sources.

the corresponding objective values are summarized in Table 9.4. Type 2 is identified as the ℓ^* type; on the other hand, none of the type-2 units should be inspected, i.e., $N_{1'}^* = 0$. For example, when demand follows the normal distribution, it is optimal to order 164 units of type 2 and 5 units of type 11, and none of them should be inspected.

9.5 Optimal Replenishment over an Infinite Horizon

We now extend the single-period model of the last section to the case of optimal replenishment over an infinite horizon, with an independent and identically distributed demand sequence, $\{D_t\}$, where D_t denotes the demand quantity in period t, with $t = 0, 1, 2, \cdots$. Any unsatisfied demand is lost, with a penalty of C_P per unit. On the other hand, for any surplus after demand is supplied, in lieu of the salvage value $s(\theta_i)$, we assume there is a holding cost $h(\theta_i)$ for each surplus unit of product i at the end of each period. Assume that $h(\theta)$, like $s(\theta)$, is a decreasing and convex function of the defective rate. Analogous to assuming that $c(\theta) - s(\theta)$ is a decreasing and convex function, here we assume that $c(\theta) - h(\theta)$ is a decreasing and convex function. This is automatically satisfied if, for example, when the holding cost is charged as a (fixed) proportion of the purchasing cost.

Furthermore, we shall assume the following two conditions:

$$C_P + h(\theta_k) \geq C_g + \Delta\theta_k + c(\theta_k) \qquad (9.21)$$

and

$$\Delta(\theta_i - \theta_{i-1}) \geq [c(\theta_{i-1}) - c(\theta_i)] - [h(\theta_{i-1}) - h(\theta_i)]. \qquad (9.22)$$

Note that (9.21) is analogous to (9.11). It guarantees that for any given type, using it to supply demand is always better than keeping the unit (and hence paying penalty and inventory charges), even if it can be salvaged at purchasing cost. Note that (9.21) is weaker than $C_P \geq C_g + \Delta\theta_k + c(\theta_k)$, which simply gives enough incentive to place orders: the shortage penalty is such that it always pays to order, including the type with the lowest quality. (Otherwise, some types can be preeliminated from the model.) And, (9.22) is analogous to (9.1): it ensures that any demand will always be supplied by the best available unit. Specifically, it is equivalent to the following:

$$C_d \theta_{i-1} + C_g(1 - \theta_{i-1}) + c(\theta_{i-1}) - h(\theta_{i-1})$$
$$\leq \quad C_d \theta_i + C_g(1 - \theta_i) + c(\theta_i) - h(\theta_i).$$

Clearly, Lemma 9.9 applies here as well. Hence, without loss of generality, we shall focus on the case of $c(0) \leq c(\theta_{\ell^*}) + C_I + C_R \theta_{l^*}$, as in Lemma 9.9, because the complementary case can be reduced to this case, as is evident from Theorem 9.11.

Let $f(N_1^{(t)}, \cdots, N_k^{(t)}; D_t)$ denote the total cost in period t, excluding the purchasing cost, provided the starting inventory *after* replenishment is $N^{(t)} := (N_1^{(t)}, \cdots, N_k^{(t)})$ and the demand is D_t. Then, following (9.17) but with $-s(\theta)$ replaced by $h(\theta)$ and with the penalty cost added, we have

$$
\begin{aligned}
& f(N_1^{(t)}, \cdots, N_k^{(t)}; D_t) \\
= \quad & C_g D_t + [C_P - C_g - \Delta \theta_k](D_t - N_{1,k}^{(t)})^+ + h(\theta_k)(D_t - N_{1,k}^{(t)})^- \\
& + \sum_{j=1}^{k} [\Delta(\theta_j - \theta_{j-1})(D_t - N_{1,j-1}^{(t)})^+ \\
& \quad + (h(\theta_{j-1}) - h(\theta_j))(D_t - N_{1,j-1}^{(t)})^-] \\
= \quad & C_g D_t + [C_P + h(\theta_k) - C_g - \Delta \theta_k](D_t - N_{1,k}^{(t)})^+ - h(\theta_k)(D_t - N_{1,k}^{(t)}) \\
& + \sum_{i=1}^{k} [\Delta(\theta_i - \theta_{i-1}) + (h(\theta_{i-1}) - h(\theta_i))](D_t - N_{1,i-1}^{(t)})^+ \\
& - \sum_{i=1}^{k} [h(\theta_{i-1}) - h(\theta_i)](D_t - N_{1,i-1}^{(t)}). \quad (9.23)
\end{aligned}
$$

Let $X^{(t)} := (X_1^{(t)}, \cdots, X_k^{(t)})$ denote the inventory level at the beginning of period t, *before* the replenishment. It is equal to the end inventory of period $t - 1$, and can be expressed as follows:

$$X_i^{(t)} = [N_i^{(t-1)} - (D_{t-1} - N_{1,i-1}^{(t-1)})^+]^+. \quad (9.24)$$

Given a replenishment policy π, denote

$$N^{\pi(t)} := (N_1^{\pi(t)}, \cdots, N_k^{\pi(t)}), \quad \text{and} \quad X^{\pi(t)} := (X_1^{\pi(t)}, \cdots, X_k^{\pi(t)}).$$

Note that for a policy π to be feasible, we must have $N_i^{\pi(t)} \geq X_i^{\pi(t)}$, for all $i = 1, \cdots, k$, and for all $t = 0, 1, \cdots$. Let $V_T^{\pi}(x)$ denote the T-period expected cost associated with the policy π, starting from

$$X^{(0)} = x := (x_1, \cdots, x_k).$$

Then we can write

$$
\begin{aligned}
&V_T^\pi(x) \\
&= \sum_{t=0}^{T-1} \mathsf{E}\{\sum_{i=1}^{k} c(\theta_i)[N_i^{\pi(t)} - X_i^{\pi(t)}] + f(N_1^{\pi(t)}, \cdots, N_k^{\pi(t)}; D_t)|X^{(0)} = x\} \\
&\quad - \sum_{i=1}^{k} c(\theta_i)\mathsf{E}[X_i^{\pi(T)}|X^{(0)} = x] \\
&= \sum_{t=0}^{T-1}\{\sum_{i=1}^{k} c(\theta_i)\mathsf{E}[N_i^{\pi(t)} - X_i^{\pi(t+1)}] + f(N_1^{\pi(t)}, \cdots, N_k^{\pi(t)}; D_t)\} \\
&\quad - \sum_{i=1}^{k} c(\theta_i)x_i
\end{aligned}
\tag{9.25}
$$

with the understanding that $X^{\pi(0)} = x$. Here the last term in (9.25) assumes that any surplus unit at the end of period T can be salvaged at purchasing cost. This term will vanish when we consider the long-run average cost.

Because $X_i^{\pi(t+1)}$ relates to $N_i^{\pi(t)}$ and D_t following (9.24), the last expression in the preceding motivates us to define:

$$
\begin{aligned}
F(N; D) \\
:= \quad & F(N_1, \cdots, N_k; D) \\
:= \quad & \sum_{i=1}^{k} c(\theta_i)\{N_i - [N_i - (D - N_{1,i-1})^+]^+\} \\
& + f(N_1, \cdots, N_k; D),
\end{aligned}
\tag{9.26}
$$

where D denotes the generic demand per period (i.e., with the same distribution as D_t). Then we can rewrite $V_T^\pi(x)$ as follows:

$$
V_T^\pi(x) = \sum_{t=0}^{T-1} \mathsf{E}[F(N^{\pi(t)}; D_t)|X^{(0)} = x] - \sum_{i=1}^{k} c(\theta_i)x_i.
$$

Denote $\bar{V}^\pi(x)$ as the long-run average cost. We have

$$
\begin{aligned}
\bar{V}^\pi(x) &= \lim_{T\to\infty} \frac{1}{T}V_T^\pi(x) \\
&= \lim_{T\to\infty} \frac{1}{T}\sum_{t=0}^{T-1} \mathsf{E}[F(N^{\pi(t)}; D_t)|X^{(0)} = x],
\end{aligned}
\tag{9.27}
$$

because $\sum_{i=1}^{k} c(\theta_i)x_i$ is a finite constant.

Lemma 9.14 The F function of (9.26) is a convex function of $(N_1, \cdots, N_k)$, and satisfies the following properties:

$$F(N_1, \cdots, N_i, \cdots, N_k; D) - F(N_1, \cdots, N_i + 1, \cdots, N_k; D)$$

is decreasing in N_j for any $j = 1, \cdots, k$; and

$$F(N_1, \cdots, N_{i-1} + 1, N_i, \cdots, N_k; D) - F(N_1, \cdots, N_{i-1}, N_i + 1, \cdots, N_k; D)$$

is increasing in $N_{1,i-1}$.

Proof. Following the proof of Proposition 9.1, part of the F function in (9.26) can be written as follows:

$$-\sum_{i=1}^{k} c(\theta_i)[N_i - (D - N_{1,i-1})^+]^+$$

$$= -\sum_{i=1}^{k}(D - N_{1,i-1})^-(c(\theta_{i-1}) - c(\theta_i)) - (D - N_{1,k})^- c(\theta_k)$$

$$= -\sum_{i=1}^{k}(D - N_{1,i-1})^+(c(\theta_{i-1}) - c(\theta_i))$$

$$+ \sum_{i=1}^{k}(D - N_{1,i-1})(c(\theta_{i-1}) - c(\theta_i)) - (D - N_{1,k})^- c(\theta_k).$$

Hence, substituting this and (9.23) into (9.26), we have

$$F(N_1, \cdots, N_k; D)$$
$$= C_g D + [C_P + h(\theta_k) - C_g - \Delta\theta_k - c(\theta_k)](D - N_{1,k})^+$$
$$+ \sum_{i=1}^{k}[\Delta(\theta_i - \theta_{i-1}) + (h(\theta_{i-1}) - h(\theta_i))$$
$$- (c(\theta_{i-1}) - c(\theta_i))](D - N_{1,i-1})^+$$
$$- \sum_{i=1}^{k}[(h(\theta_{i-1}) - h(\theta_i)) - (c(\theta_{i-1}) - c(\theta_i))](D - N_{1,i-1})$$
$$- (h(\theta_k) - c(\theta_k))(D - N_{1,k}) + \sum_{i=1}^{k} c(\theta_i)N_i.$$

Note that on the right-hand side, both $(D - N_{1,i-1})^+$ and $(D - N_{1,k})^+$ are convex in $(N_1, ..., N_k)$ and their coefficients are nonnegative, following (9.21) and (9.22). The other terms are all linear in $(N_1, ..., N_k)$. Hence, F is convex in $(N_1, \cdots, N_k)$.

The other two properties follow immediately from the close resemblance of $F(N_1, \cdots, N_k; D)$ to the V function in (9.12). $\square$

In view of Lemma 9.14, the minimizer

$$N^* := (N_1^*, \cdots, N_k^*) := \arg \min \mathsf{E}F(N_1, ..., N_k; D) \qquad (9.28)$$

is well defined. Also, define

$$G_i^*(N_{1,i})$$
$$:= \mathsf{E}F(N_1, \cdots, N_{i-1}+1, N_i, \cdots, N_k; D)$$
$$-\mathsf{E}F(N_1, \cdots, N_{i-1}, N_i+1, \cdots, N_k; D).$$

Then, analogous to (9.19), $G_i^*(N_{1,i}) > 0$ is equivalent to the following inequality:

$$\frac{h(\theta_i) - h(\theta_{i+1})}{\theta_{i+1} - \theta_i}$$
$$> [\Delta - \frac{(c(\theta_i) - h(\theta_i)) - (c(\theta_{i+1}) - h(\theta_{i+1}))}{\theta_{i+1} - \theta_i}]$$
$$\cdot [\mathsf{E}(D - N_{1,i})^+ - \mathsf{E}(D - N_{1,i} - 1)^+]. \qquad (9.29)$$

Because $h(\theta)$ is decreasing and convex, the left-hand side of (9.29) is decreasing in i. Similarly, because $c(\theta) - h(\theta)$ is decreasing and convex, the first factor on the right-hand side of (9.29) is increasing in i. Hence, the discussion preceding Theorem 9.10 and the results stated in Theorem 9.10 apply here as well. In particular, the optimal solution in (9.28) can be generated by the greedy algorithm in the last section, with (9.29) replacing (9.19).

Denote the set of vectors,

$$S := \{x : x_i \le N_i^* \text{ for } 1 \le i \le k \},$$

with N^* being the minimizer in (9.28). Without loss of generality, we shall assume $\mathsf{E}[D] > 0$ (otherwise, we have the trivial case of $D \equiv 0$). We are now ready to study the optimal policy that minimizes the long-run average cost objective in (9.27). Theorem 9.15 states that the optimal policy is to order up to the level $N^* = (N_i^*)$, unless the inventory (of any type, in any period) already exceeds this level, in which case you should order nothing.

Theorem 9.15 It is optimal to order up to N^* in period t, whenever $X^{(t)} \in S$ and to order nothing if $X^{(t)} \notin S$.

Proof of Theorem 9.15. From (9.27), we have

$$\bar{V}^\pi(x) = \lim_{T \to \infty} \frac{1}{T} \sum_{t=0}^{T-1} \mathsf{E}[F(N^{\pi(t)}; D_t)|X^{(0)} = x]$$

$$\ge \lim_{T \to \infty} \frac{1}{T} \sum_{t=0}^{T-1} \mathsf{E}F(N^*; D_t)$$

$$= \mathsf{E}F(N^*; D). \qquad (9.30)$$

We will show that $\bar{V}^{\pi^*}(x) = \mathsf{E}F(N^*; D)$ for any initial inventory x, so that π^* is optimal. (The feasibility of π^* is obvious.)

First, note that if $x = X^{(0)} \in S$, then $X^{\pi^*}(t) \in S$ and hence $N^{\pi^*}(t) = N^*$ for all $t \geq 0$. Therefore,

$$\bar{V}^{\pi^*}(x) = \lim_{T \to \infty} \frac{1}{T} \sum_{t=0}^{T-1} \mathsf{E}[F(N^{\pi^*}(t); D_t)|X^{(0)} = x]$$

$$= \lim_{T \to \infty} \frac{1}{T} \sum_{t=0}^{T-1} \mathsf{E}F(N^*; D_t)$$

$$= \mathsf{E}F(N^*; D).$$

Now suppose $x \notin S$. Define

$$T_x := \inf\{0 \leq t < \infty : X^{\pi^*}(t) \in S, X^{(0)} = x\},$$

i.e., T_x is the time until the inventory level drops down the set S. Note that under the stated policy π^*, nothing will be ordered until T_x. Hence,

$$T_x \leq \hat{T}_x := \min\{T : \sum_{t=0}^{T} D_t \geq \sum_{i=1}^{k} x_i\},$$

and we must have $\mathsf{E}[\hat{T}_x] < \infty$, because $\mathsf{E}[D] > 0$; and hence, $\mathsf{E}[T_x] < \infty$. Furthermore, from (9.23) and (9.26), it is clear that, for each period $t < T_x$, the expected cost $\mathsf{E}[F]$ is bounded:

$$0 \leq \mathsf{E}[F(N^{\pi^*}(t); D_t)|X^{(0)} = x] \leq B_x$$

for some constant B_x (which may depend on x). Hence, when $x \notin S$, we have

$$\bar{V}^{\pi^*}(x) = \lim_{T \to \infty} \frac{1}{T} \sum_{t=0}^{T-1} \mathsf{E}[F(N^{\pi^*}(t); D_t)|X^{(0)} = x]$$

$$\leq \lim_{T \to \infty} \frac{1}{T}\{B_x \mathsf{E}[T_x] + \sum_{t=T_x}^{T-1} \mathsf{E}[F(N^{\pi^*}(t); D_t)|X^{(0)} = x]$$

$$= \lim_{T \to \infty} \frac{1}{T} \sum_{t=T_x}^{T-1} \mathsf{E}F(N^*; D_t)$$

$$= \mathsf{E}F(N^*; D).$$

Combining this with (9.30), we have $\bar{V}^{\pi^*}(x) = \mathsf{E}F(N^*; D)$. $\square$

Note that when the initial inventory exceeds the desired level of N^*, nothing is ordered and the inventory level will be brought down to below N^* within a finite time, during which the expected one-step cost is bounded.

Type i	1	2	3	Obj. Val.
Defective rate θ_i	.00	.01	.03	
(N_1^*, N_2^*, N_3^*) (N)	151	3	24	623.0
(N_1^*, N_2^*, N_3^*) (U)	151	3	24	616.9
(N_1^*, N_2^*, N_3^*) (P)	150	2	15	613.3

TABLE 9.5. Infinite horizon; convex $c(\theta)$ and $h(\theta)$.

This is guaranteed by the fact that it always pays to use up any unit of inventory to supply demand instead of keeping it, thanks to the assumed condition in (9.21). Hence, any cost over this finite time will be washed out in the long-run average. This is in contrast to the models in Ignall and Veinott [46] and Veinott [99], where the initial inventory must be restricted to below N^*. Two aspects of those models are different from our model here: (a) multiple demand types, with the possibility of substitution (whereas we only consider a single demand stream); and (b) allowing backlog (we assume lost sales). Hence, in those models it is possible that the initial inventory of some types may be kept forever, while some other types may run into a large amount of backlog if nothing is ordered. This will result in an unbounded one-step cost, and the argument in our proof will not apply.

In summary, finding the optimal replenishment policy in the infinite horizon case amounts to solving a single-period problem, exactly like the one in the last section; in particular, the optimal order-up-to level, N^*, can be derived from the greedy algorithm there. Furthermore, when both $c(\theta)$ and $h(\theta)$ are linear, in the infinite-horizon case we also have the optimality of dual sourcing (from sources 1 and k), just as in the single-period case.

Example 9.16 Continue with the problem in Example 9.13, but with an infinite horizon, and three product types only (with slightly different defective rates). There is a perfect type, indexed as $i = 1$; and the ordering cost, $c(\theta) = 3.0 - 5\theta + 6\theta^2$, applies to all three types. Other data remain the same as before. In addition, assume the holding cost is 20% of the purchasing cost: $h(\theta) = 0.2c(\theta)$. Note that here we have $c(0) < c(\theta_i) + C_I + C_R\theta_i$, for $i = 2, 3$; and consequently, no inspection of any type is performed.

For the three types of demand distributions in Example 9.13, the optimal order-up-to quantities and the corresponding objective values are summarized in Table 9.5.

Table 9.6 repeats the results, but with both the purchasing and holding costs being linear in θ: $c(\theta) = 3.0 - 5\theta$ and $h(\theta) = 0.2c(\theta)$. Note that $N_2^* = 0$ in all cases, as expected, because dual sourcing is optimal.

Type i	1	2	3	Obj. Val.
defective rate θ_i	.00	.01	.03	
(N_1^*, N_2^*, N_3^*) (N)	150	0	28	623.0
(N_1^*, N_2^*, N_3^*) (U)	149	0	29	616.8
(N_1^*, N_2^*, N_3^*) (P)	150	0	17	613.2

TABLE 9.6. Infinite horizon; linear $c(\theta)$ and $h(\theta)$

9.6   A Random Yield Model with Multiple Sources

We can recast the model studied earlier as a random yield model (refer to, e.g., [2, 36, 54, 67, 94, 105]) as follows. For ease of discussion, we focus on the single-period case. It would thus be helpful to relate to the model in §9.4.

Suppose there are k sources of supply, indexed by $i = 1, ..., k$. Each source i has a yield ratio of $1 - \theta_i$. Specifically, a proportion, θ_i, of any quantity ordered from source i may not be delivered; in other words, an order of N_i units will result in an expected delivery of $N_i(1 - \theta_i)$ units—the actual "yield" that can be used to supply demand. Assume, as before, that the sources are indexed in increasing order of the θ_i values.

Let D denote demand as before. Reinterpret C_g and C_d as the costs for each unit of demand satisfied and unsatisfied. In particular, $-C_g$ is the profit derived from supplying each unit of demand. As before, let $\Delta = C_d - C_g$. Note that with $C_d \geq 0$ and $-C_g \geq 0$, $\Delta \geq 0$ is automatic. Let $s \geq 0$ denote the salvage value for each surplus unit; we replace condition (9.1) with $\Delta \geq s$. Note that because the sources only differ in their yield ratios, once delivered all units are of equal value in supplying demand, hence, the salvage value is independent of the sources, just like C_g and C_d. Also note that under the new interpretation of C_d, it is necessarily equal to the penalty cost in §9.4, C_P; hence, the inequality in (9.11) is automatically satisfied. Let $c(\theta_i)$ be the purchasing cost of each unit from source i. For example, $c(\theta_i) = c(1 - \theta_i)$, where $c > 0$ is the cost rate for each unit delivered; hence, $cN_i(1 - \theta_i)$ is the expected purchasing cost for an order quantity of N_i units from source i. As before, assume that $c(\theta_i)$ is a decreasing and convex function.

Our model allows a new feature not present in previous random yield models. At a premium—above and beyond the purchasing cost—of $a + b\theta_i$ per unit, the supply (delivery) can be guaranteed. Hence, out of the N_i units ordered from source i, for which we pay a purchasing cost of $N_i c(\theta_i)$, we may choose to guarantee a delivery of n_i units by paying an additional premium of $n_i(a + b\theta_i)$. Clearly, this feature is analogous to upgrading, through inspection and repair, a defective unit to a perfect unit. Hence, the (per-unit) premium, $a + b\theta_i$, corresponds to the inspection and repair costs in the earlier model with $a = C_I$ and $b = C_R$. To facilitate comparisons, we shall continue writing C_I and C_R instead of a and b.

We now illustrate how the model in §9.4 can be adapted to identify the optimal order quantities from a set of k unreliable supply sources, each having a random yield in quantities actually delivered. The replenishment decision is supplemented by the option of paying a premium to secure a guaranteed delivery quantity (which is analogous to inspection in the earlier model); the objective is to minimize the expected total net cost—purchasing, premium, and penalty costs minus profit and salvage value.

We start with the expression in (9.2) for $W(D, n_1, \cdots, n_k)$, reinterpreted here as the penalty cost minus profit and salvage value, given that the demand is D, the replenishment quantities are $(N_1, ..., N_k)$, of which $(n_1, \cdots, n_k)$ are guaranteed by paying premiums. Let $B_i(N_i - n_i)$ denote the number of units from source i that are not guaranteed by premiums but are actually delivered. We shall assume that $B_i(N_i - n_i)$ follows a binomial distribution associated with $N_i - n_i$ Bernoulli trials, each with a success probability of $1 - \theta_i$. This implies that the yield of each unit is independent of all other units. This independence assumption is crucial to the following stochastic monotonicity results.

As before, denote $n_{1,k} = \sum_{i=1}^{k} n_i$; similarly, denote

$$B_{1,k} = \sum_{j=1}^{k} B_j(N_j - n_j)$$

and

$$B_{1,k}^i = \sum_{\substack{j=1 \\ j \neq i}}^{k} B_j(N_j - n_j) + B_i(N_i - n_i - 1).$$

Then we can rewrite the function W in (9.2) as follows:

$$
\begin{aligned}
& W(D, n_1, \cdots, n_k) \\
=~& C_g \min\{n_{1,k} + B_{1,k}, D\} + C_d(D - n_{1,k} - B_{1,k})^+ \\
& \quad -s(n_{1,k} + B_{1,k} - D)^+ \\
=~& C_g D + \Delta(D - n_{1,k} - B_{1,k})^+ - s(n_{1,k} + B_{1,k} - D)^+ \\
=~& C_g D + (\Delta - s)(D - n_{1,k} - B_{1,k})^+ \\
& \quad +s(D - n_{1,k} - B_{1,k}). \quad\quad\quad (9.31)
\end{aligned}
$$

The expression for Π in (9.3) remains valid here, with the new interpretation of $C_I + C_R\theta_i$ being the premium to guarantee the delivery of a unit from source i. Therefore, (9.5) becomes:

$$
\begin{aligned}
& \Pi(n_1, \cdots, n_{i-1}, n_i, \cdots, n_k) - \Pi(n_1, \cdots, n_{i-1}, n_i + 1, \cdots, n_k) \\
=~& -(C_I + C_R\theta_i) + s\theta_i + (\Delta - s) \\
& \quad \cdot [\mathsf{E}(D - n_{1,k} - B_{1,k})^+ - \mathsf{E}(D - n_{1,k} - B_{1,k}^i - 1)^+]. \quad (9.32)
\end{aligned}
$$

Note that the second expectation follows from the fact that when n_i is increased to $n_i + 1$, $n_{1,k}$ becomes $n_{1,k} + 1$ and $B_{1,k}$ becomes $B^i_{1,k}$.

Note the following relation:

$$B_{1,k} = B^i_{1,k} + \delta_i,$$

where δ_i denotes a binary variate that equals 1 with probability $1 - \theta_i$. We can modify (9.4) based on (9.32):

$$
\begin{aligned}
&\Pi(n_1, \cdots, n_{i-1} + 1, n_i, \cdots, n_k) - \Pi(n_1, \cdots, n_{i-1}, n_i + 1, \cdots, n_k)\\
={}& -(C_R - s)(\theta_i - \theta_{i-1}) + (\Delta - s)\\
&\cdot[\mathsf{E}(D - n_{1,k} - B^{i-1}_{1,k} - 1)^+ - \mathsf{E}(D - n_{1,k} - B^i_{1,k} - 1)^+]\\
={}& -(C_R - s)(\theta_i - \theta_{i-1}) + (\Delta - s)\\
&\cdot[\mathsf{E}(D - n_{1,k} - B_{1,k} + \delta_{i-1} - 1)^+ - \mathsf{E}(D - n_{1,k} - B^i_{1,k} + \delta_i - 1)^+]\\
={}& -(C_R - s)(\theta_i - \theta_{i-1}) + (\Delta - s)(\theta_i - \theta_{i-1})\\
&\cdot[\mathsf{E}(D - n_{1,k} - B_{1,k})^+ - \mathsf{E}(D - n_{1,k} - B_{1,k} - 1)^+].
\end{aligned}
\tag{9.33}
$$

With the same definition of $g_i(n_{i,k})$ as in §9.2, here $g_i(n_{i,k}) < 0$ is equivalent to

$$C_R - s > (\Delta - s)\mathsf{E}[(D - y)^+ - (D - y - 1)^+] \tag{9.34}$$

with $y := n_{1,k} + B_{1,k}$. Note that the right-hand side of (9.34) is decreasing in n_i for any $i = 1, ..., k$. This fact follows from the standard theory of stochastic monotonicity (Ross [73], chapter 9), because (a) $(D - x - 1)^+ - (D - x - 2)^+$ is a decreasing function of x, and (b) $n_i + B_j(N_i - n_i)$ is stochastically increasing in n_i. Furthermore, like the case of (9.8) and (9.9), the right-hand side of (9.34) decreases as we increase n_i, in the increasing order of i. (This follows from the fact that $B^i_{1,k}$ is stochastically increasing in i.) Similarly, the right-hand side of (9.32) is also decreasing in n_i, for any $i = 1, ..., k$.

Therefore, with the order quantities, $(N_1, ..., N_k)$, given, the decision problem of finding the optimal number of units to guarantee from each source through paying a premium has the same structure as the optimal inspection problem in §9.3. In particular, with (9.5) and (9.8) replaced by (9.32) and (9.34), and condition (9.10) replaced by $\Delta = s = C_R$, all the results in Theorem 9.6 are still applicable here. In particular, the optimal solution here should be $(N_1, ..., N_{j^*-1}, n'_{j^*}, 0, ..., 0)$, where $j^* \geq 1$ is the smallest index, and $n'_{j^*} \leq N_{j^*}$ is the smallest n_i (with $i = j^*$) value, such that the cost reduction in (9.32) becomes nonpositive; if no such value exists, then $j^* = k$ and $n'_{j^*} = N_k$.

Now consider the order sizes as decision variables as well. Lemma 9.9 is still valid here, and thus so is the expression in (9.14) for V, but without the last term, the penalty cost, which is now part of W. Note that when no premium is paid to guarantee delivery, the first two terms on the right-hand side of both (9.32) and (9.33) vanish. Consequently, we have

$$
\begin{aligned}
& V(N_1, \cdots, N_i, \cdots, N_k) - V(N_1, \cdots, N_i + 1, \cdots, N_k) \\
= {} & -c(\theta_i) + (\Delta - s)(\theta_i - \theta_{i-1}) \\
& \cdot [\mathsf{E}(D - B_{1,k})^+ - \mathsf{E}(D - B_{1,k} - 1)^+]
\end{aligned}
\tag{9.35}
$$

and

$$
\begin{aligned}
& V(N_1, \cdots, N_{i-1} + 1, N_i, \cdots, N_k) - V(N_1, \cdots, N_{i-1}, N_i + 1, \cdots, N_k) \\
= {} & c(\theta_{i-1}) - c(\theta_i) - (\Delta - s)(\theta_i - \theta_{i-1}) \\
& \cdot [\mathsf{E}(D - B_{1,k})^+ - \mathsf{E}(D - B_{1,k} - 1)^+].
\end{aligned}
\tag{9.36}
$$

Note that in these expressions, $B_{1,k} := B_{1,k}(N_{1,k})$, which is stochastically increasing in N_i for any $i = 1, ..., k$. Hence, the right-hand sides of (9.35) and (9.36) are, respectively, decreasing and increasing in N_i for any $i = 1, ..., k$, just like (9.15) and (9.16). In particular, (9.19) now becomes

$$
\begin{aligned}
& \frac{c(\theta_i) - c(\theta_{i+1})}{\theta_{i+1} - \theta_i} \\
> {} & (\Delta - s)[\mathsf{E}(D - B_{1,k}(N_{1,i}))^+ - \mathsf{E}(D - B_{1,k}(N_{1,i}) - 1)^+],
\end{aligned}
\tag{9.37}
$$

where we have explicitly written the argument of $B_{1,k}$ in correspondence to the i index on the left-hand side. This way, as in §9.4, for each i, we can increase N_i until the inequality is satisfied. On the other hand, every time i is increased, the left-hand side decreases, requiring an increase in the $N_{1,i}$ value to satisfy the inequality in (9.37). This leads to the sequence of M_i values, with $M_i - M_{i-1}$ being the upper limit on the order size from source i. On the other hand, the decreasing property of the right-hand side of (9.35) plays the role of a stopping rule, exactly as in §9.4.

Therefore, the solution to the optimal replenishment problem can be obtained following the two theorems in §9.4: we first compare the purchasing cost of the perfectly reliable supply source, $i = 1$, with the supply source ℓ^* that has the lowest combined purchasing cost and premium among all other sources. If source 1 is less expensive, then the optimal order quantities are obtained in increasing order of i, starting from $i = 1$ and following the upper limits specified by the M_is and the stopping rule signified by the nonpositive cost reduction in (9.15); no premium should be paid to any sources. Otherwise, ignore source 1, replace it by source ℓ^* with premiums paid for all units ordered to guarantee a perfect yield; and then proceed in the same manner as in the previous case. Extensions to the infinite horizon as in §9.5 lead to the optimality of the order-up-to policy.

9.7 Notes

It is quite common for production-inventory systems to have multiple supply sources that have different grades of reliability in terms of the quantity and quality of orders delivered; refer to, e.g., Anupindi and Akella [2], and Parlar and Wang [67]; also refer to Chen, Yao and Zheng [20], from which most of the material in this chapter was drawn.

In this kind of setting, it is imperative that replenishment decisions take into account supply uncertainty and related cost implications, in addition to the usual tradeoff between the possibilities of surplus inventory and unmet demand.

Along with replenishment decisions, there are recourse actions that can be taken to offset supply imperfection. For example, certain quality control mechanisms can be applied to the orders received, including rework on any defective units before the orders are supplied to customers. Here, the quality control decision (on inspection and repair) is embedded into the replenishment decision; typically both decisions have to be made before demand is realized. The existing literature in this area focuses mostly on a single unreliable supply source; refer to, for example, Lee [52], Lee and Rosenblatt [53], Peter, Schneider, and Tang [70], and Yao and Zheng [109], among others.

Another kind of recourse action to offset yield loss is "substitution," which uses the surplus of higher-grade products (in terms of quality and functionality, for example) to supply the shortage of demand for lower-grade products. Unlike the inspection-repair mechanism, substitution cannot be carried out until the demand is realized. On the other hand, like the inspection-repair mechanism, which is tantamount to paying premiums to offset yield loss, substitution incurs extra costs associated with filling demand for a lower-grade product with a higher-grade product. Substitution is the subject of the next chapter.

10

Inventory Control with Substitution

The focus of this chapter is on substitutable inventory systems, which, as we pointed out at the end of the last chapter (as well as in the introductory chapter), complements quality inspection in terms of providing additional means of recourse *after* demand is realized, as the surplus of higher-end products can be used to substitute for lesser products. There are two interleaved decisions. The main decision is the order quantity for each product type at the beginning of the period (i.e., before demands are realized). This, however, has to take into account the substitution scheme, which also needs to be carried out in an optimal fashion, at the end of the period when demands are realized.

We start with model formulation in §10.1 and develop the optimal substitution policy in §10.2. The optimal replenishment problem is then formulated in §10.3, with the properties of the objective function, such as concavity and submodularity, studied in §10.4. These properties lead to the optimal solution to the order quantities in §10.5 and their bounds in §10.6.

10.1 Model Description

There are N product types, indexed by the subscript $n = 1, ..., N$. Alternatively, we shall use indices i, j, k, etc. A product with a higher rank (1 being the highest; N the lowest) in general has better quality and more functionality and hence can substitute for a product in a lower rank if necessary. More details along this line will follow.

The planning horizon is a single period. Each type n has its own demand over the period, D_n; and $\{D_n, n = 1, ..., N\}$ is a set of independent random variables. At the beginning of the period, the demand for each type is only known through its distribution. The actual demand is realized at the end of the period.

There are two decisions that need to be carried out over the period. The replenishment decision—how many units to order for each product type—has to be made at the beginning of the period. The substitution decision is made at the end of the period when demands are realized. This concerns whether to use the surplus of some product type to substitute for other types that experience a shortage after supplying demands (provided the substitution is feasible—to be specified later). It is important to note here that the replenishment decision at the beginning of the period has to take into account the substitution decision at the end of the period.

At the beginning of the period, there are $x_n \geq 0$ units of on-hand inventory for each type n. The decision is to place an order to bring the inventory level to $y_n \geq x_n$ for each type, knowing only the distribution of D_n. Ignore delivery lead time so that the order arrives immediately. Demands are then realized, denoted d_n for type n.

Other data are as follows. For each unit of type n product, c_n and p_n are the acquisition cost and the selling price; h_n and q_n are the inventory holding cost and the shortage penalty; and s_n is the salvage value for any surplus at the end of the period. Note that, although it is not necessary to include a holding cost for a single-period problem, it will be more meaningful when we discuss the multiperiod problem later.

We focus on a "downward substitution" rule, which allows supplying demand for type-j products using type-i products, for any $i \leq j$. Note, however, that if a unit of type i supplies the demand for type j, the price charged is p_j (instead of p_i). Also note that we may choose not to supply demand for type j using type i, even if there is a shortage of j and a surplus of i. This will be examined more closely later.

For convenience, define

$$r_n := p_n + q_n \qquad \text{and} \qquad v_n := s_n - h_n$$

for $n = 1, ..., N$. We need the following conditions on the cost data involved.

Condition 10.1   (i) $r_n \geq v_n$ and $r_n \geq c_n$;

(ii) r_n and v_n are both decreasing in n.

Note that (i) is basically a regularity condition: it ensures that each type of products will indeed be used to supply demand (for that type) instead of being held as inventory and exchanged for salvage value (substitution for another type, of course, is another matter); and that there is incentive for placing orders (otherwise, "do nothing" could be trivially optimal). The condition in (ii) essentially associates more "weight" to product types with higher ranks (which can substitute for those of lower ranks). It will play a key role in the problem structure, as we shall demonstrate.

Denote the following vectors:

$\mathbf{x} := (x_n)_{n=1}^N$: initial on-hand inventory levels, before placing the order;

$\mathbf{y} := (y_n)_{n=1}^N$: on-hand inventory levels, after receiving the order;

$\mathbf{D} := (D_n)_{n=1}^N$: demands (random);

$\mathbf{d} := (d_n)_{n=1}^N$: realized demands.

The partial ordering between two vectors, $\mathbf{x} \leq \mathbf{y}$, will refer to the usual componentwise ordering. We shall use $\mathbf{x} < \mathbf{y}$ to denote $\mathbf{x} \leq \mathbf{y}$ and $\mathbf{x} \neq \mathbf{y}$ (hence, $\mathbf{y}$ has at least one component that is strictly larger than the corresponding component of $\mathbf{x}$); and $\mathbf{x} << \mathbf{y}$ to denote $x_i < y_i$ for all components i. We shall also denote $[x]^+ := \max(x, 0)$. Throughout, the terms, 'increasing', 'decreasing', 'convex' and 'concave' are used in the nonstrict sense.

10.2   The Optimal Substitution Policy

Suppose now that the replenishment order has arrived (so that the on-hand inventory has been brought up to $\mathbf{y}$) and demands are realized ($\mathbf{d}$). First we want to decide how to use the inventories to supply the demands, allowing substitution. Specifically, let $w_{i,j}$ be the number of type i units allocated to supply type j demand, for all $j \geq i$, following the downward substitution rule. Let $\mathbf{w} := (w_{i,j})_{i \leq j}$. We want to find an allocation that solves the following maximization problem:

$$\max_{\mathbf{w}} \sum_{j=1}^N \{ p_j \min (d_j, w_{j,j} + \sum_{i=1}^{j-1} w_{i,j})$$

$$+ v_j (y_j - w_{j,j} - \sum_{k=j+1}^N w_{j,k}) - q_j [d_j - w_{j,j} - \sum_{i=1}^{j-1} w_{i,j}]^+ \},$$

$$\text{s.t.} \quad w_{j,j} + \sum_{k=j+1}^N w_{j,k} \leq y_j, \quad j = 1, ..., N;$$

$$w_{i,j} \geq 0, \quad i = 1, ..., N; \ j \geq i.$$

The first term in the objective function is the revenue (selling price) from supplying the demands, both directly and using substitution. The second term is the net salvage value (i.e., after subtracting inventory holding cost) for excess inventory. The third term is the penalty cost for shortage. Notice that

$$(d_j - w_{j,j} - \sum_{i=1}^{j-1} w_{i,j})^+$$

$$= \max(d_j, w_{j,j} + \sum_{i=1}^{j-1} w_{i,j}) - w_{j,j} - \sum_{i=1}^{j-1} w_{i,j}$$

$$= -\min(d_j, w_{j,j} + \sum_{i=1}^{j-1} w_{i,j}) + d_j,$$

where the first equality is from the fact that

$$\max(a, b) = (a - b)^+ + b$$

and the second equality is due to the identity of

$$\min(a, b) + \max(a, b) = a + b.$$

We can rewrite the maximization problem as follows:

$$\max_{\mathbf{w}} \quad \sum_{j=1}^{N} \{ r_j \min(d_j, w_{j,j} + \sum_{i=1}^{j-1} w_{i,j})$$

$$+ v_j (y_j - w_{j,j} - \sum_{k=j+1}^{N} w_{j,k}) \} \tag{10.1}$$

$$\text{s.t.} \quad w_{j,j} + \sum_{k=j+1}^{N} w_{j,k} \le y_j, \quad j = 1, ..., N; \tag{10.2}$$

$$w_{i,j} \ge 0, \quad i = 1, ..., N; \; j \ge i.$$

Note that we have omitted the term $-\sum_{j=1}^{N} q_j d_j$ from the objective function, because it is independent of the decision variables.

Proposition 10.2 There exists an optimal substitution policy with the following structure:

(i) always supply demand for type j using on-hand type j inventory as much as possible, i.e., $w_{j,j} = \min(y_j, d_j)$, for $j = 1, ..., N$;

(ii) if there is excess inventory for type j, i.e. $y_j > d_j$, then use it to substitute for type $k > j$ shortage: $y_k < d_k$, in ascending order (i.e, $k = j + 1, j + 2, ...$), and only if $v_j \le r_k$.

(iii) if there is a shortage of type j, i.e., $y_j < d_j$, then use substitution from type $k < j$ surplus: $y_k > d_k$, in descending order (i.e., $k = j-1, j-2, ...$), and only if $v_k < r_j$.

Proof. (i) Suppose $w_{j,j} < \min(y_j, d_j)$. Then we can increase $w_{j,j}$ by one unit. There are two cases:

(a) The constraint in (10.2) is still satisfied. Then we will have a positive increase in the second term in (10.1), without decreasing the first term. Hence, either this case cannot happen or $\mathbf{w}$ is not optimal.

(b) The constraint in (10.2) is violated. This implies $w_{j,k} > 0$ for some $k > j$. Hence, reduce this $w_{j,k}$ by one unit to maintain feasibility. (Note that before and after the change, the second term in (10.1) stays zero.) This way, we increase the objective value by r_j and decrease it by r_k. Because $r_j \geq r_k$ following Assumption 10.1 (i), the objective value after making the changes is at least as good as the one achieved by $\mathbf{w}$.

(ii) To be specific and without loss of generality, suppose $y_k < d_k$ for $k = j+1$ and $j+2$. Then substituting one unit of type j for type $j+1$ (i.e., increasing $w_{j,j+1}$ by one unit) incurs a net change of the objective value by $r_{j+1} - v_j$, whereas substituting for type $j+2$ incurs a net change of $r_{j+2} - v_j$. Because $r_{j+1} \geq r_{j+2}$, following Assumption 10.1 (ii), it is more desirable to substitute for $j+1$. On the other hand, if $r_{j+1} - v_j < 0$, then even the substitution for $j+1$ will not be worthwhile.

(iii) Similar to the argument in (ii), without loss of generality, suppose $y_k > d_k$ for $k = j-1$ and $j-2$. Then substituting one unit of type $j-1$ for type j (i.e., increasing $w_{j-1,j}$ by one unit) incurs a net change of the objective value by $r_j - v_{j-1}$, whereas using type $j-2$ incurs a net change of $r_j - v_{j-2}$. Because $v_{j-2} \geq v_{j-1}$, following Assumption 10.1 (ii), it is more desirable to use $j-1$. On the other hand, if $r_j - v_{j-1} < 0$, then even using type $j-1$ for j will not be worthwhile. $\square$

Based on the substitution policy specified in this proposition, we shall use the statement, "type i can substitute for type j" $(i < j)$, denoted $i \to j$, to mean that $v_i \leq r_j$, i.e., part (ii) of the substitution policy in Proposition 10.2 is satisfied, and hence type i can *profitably* substitute for type j. Note that $i \to k$ implies $i \to j$ for $i < j < k$, because $v_i \leq r_k \leq r_j$, following the decreasing property of r_n in Assumption 10.1 (ii). Similarly, $i \to k$ implies $j \to k$ for $i < j < k$, because $v_j \leq v_i \leq r_k$, following the decreasing property of v_n. Hence, we can define

$$a(i) = \max\{j \geq i : i \to j\} \quad \text{and} \quad a^{-1}(i) = \min\{j \leq i : j \to i\}. \quad (10.3)$$

That is, $a(i)$ denotes the lowest ranked product type that i can profitably substitute for, while $a^{-1}(i)$ denotes the highest ranked product type that can profitably substitute for i.

Lemma 10.3 (i) $a(i) \leq a(j)$ and $a^{-1}(i) \leq a^{-1}(j)$ for $i < j$. That is, both $a(i)$ and $a^{-1}(i)$ are increasing in i.
(ii) $a(i) \geq a(a^{-1}(i)) \geq i$ and $a^{-1}(i) \leq a^{-1}(a(i)) \leq i$.

Proof. (i) Suppose $a(i) = i^*$. For $j > i$, if $i^* \leq j$, then

$$a(j) \geq j \geq i^* = a(i).$$

If $i^* > j$, then $i \to i^*$ implies $j \to i^*$. Hence, $a(j) \geq i^*$ follows from the fact that

$$a(j) = \max\{k \geq j : j \to k\}.$$

The increasingness of a^{-1} is similar.
    (ii) Because $a^{-1}(i) \leq i$, the increasing property in (i) implies

$$a(i) \geq a(a^{-1}(i)).$$

Meanwhile, suppose $a^{-1}(i) = j^*$. Then $j^* \to i$, and hence

$$a(j^*) = \max\{k \geq j : j^* \to k\} \geq i.$$

The same reasoning holds for the other pair of inequalities. $\square$

10.3 Formulation of the Replenishment Decision

Now consider a set of product types i through j, with $i \leq j \leq a(i)$ (i.e., i is the highest ranked type and j is the lowest). Let $S_{i,j}$ and $H_{i,j}$ denote, respectively, the total shortage and total inventory of product types i through j, $1 \leq i \leq j \leq N$, following the optimal substitution rule of Proposition 10.2. For convenience, we also define $S_{k,j} = H_{k,j} = 0$ for all $k > j$.

Lemma 10.4 For $i \leq j \leq a(i)$, we have

$$S_{i,j} = (S_{i+1,j} + d_i - y_i)^+ \quad \text{and} \quad H_{i,j} = (H_{i,j-1} + y_j - d_j)^+.$$

Proof. It is trivial for $i = j$, as $S_{j+1,j} = H_{j+1,j} = 0$. Suppose $i < j$. If $y_i < d_i$, then the total shortage is $S_{i+1,j} + d_i - y_i$ because no other type can substitute for i. On the other hand, if $y_i \geq d_i$, then the difference, $y_i - d_i$ can be used to reduce the shortage $S_{i+1,j}$, as $j \leq a(i)$ and hence $i \to j$. The other recursion is similarly argued. $\square$

    To lighten notation, we will denote

$$y_{i,j} := y_i + \cdots + y_j \quad \text{and} \quad d_{i,j} := d_i + \cdots + d_j$$

with the understanding that $y_{i,j} = d_{i,j} = 0$ whenever $i > j$.

Lemma 10.5 For $i \leq j \leq a(i)$, we have

$$
\begin{aligned}
& (y_i - d_i - S_{i+1,j})^+ \\
= \quad & \max\left(y_{i,j} - d_{i,j}, y_{i+1,j} - d_{i+1,j}, \cdots, y_j - d_j, 0\right) \\
& - \max\left(y_{i+1,j} - d_{i+1,j}, y_{i+2,j} - d_{i+2,j}, \cdots, y_j - d_j, 0\right) \\
:= \quad & M(i,j) - M(i+1,j)
\end{aligned}
\tag{10.4}
$$

and

$$
\begin{aligned}
& (d_j - y_j - H_{i,j-1})^+ \\
= \quad & \max\left(d_{i,j} - y_{i,j}, d_{i,j-1} - d_{i,j-1}, \cdots, d_i - d_i, 0\right) \\
& - \max\left(d_{i,j-1} - y_{i,j-1}, d_{i,j-2} - y_{i,j-2}, \cdots, d_i - y_i, 0\right) \\
= \quad & M(i,j) - M(i,j-1) - (y_i - d_i).
\end{aligned}
\tag{10.5}
$$

Proof. From Lemma 10.4, we have

$$
(y_i - d_i - S_{i+1,j})^+ = \max\left(y_i - d_i, S_{i+1,j}\right) - S_{i+1,j}.
\tag{10.6}
$$

On the other hand,

$$
\begin{aligned}
S_{i+1,j} & = (d_{i+1} - y_{i+1} + S_{i+2,j})^+ \\
& = [S_{i+2,j} - (y_{i+1} - d_{i+1})]^+ \\
& = \max\left(y_{i+1} - d_{i+1}, S_{i+2,j}\right) + d_{i+1} - y_{i+1}.
\end{aligned}
\tag{10.7}
$$

Subtracting $d_{i+1} - y_{i+1}$ from both terms on the right-hand side of (10.6) and taking into account (10.7), we have

$$
\begin{aligned}
& (y_i - d_i - S_{i+1,j})^+ \\
= \quad & \max\left(y_{i,i+1} - d_{i,i+1}, y_{i+1} - d_{i+1}, S_{i+2,j}\right) \\
& - \max\left(y_{i+1} - d_{i+1}, S_{i+2,j}\right).
\end{aligned}
$$

Repeating this procedure leads to (10.4). The recursion in (10.5) is similarly derived, replacing (10.6) and (10.7) by

$$
(d_j - y_j - H_{i,j-1})^+ = \max\left(d_j - y_j, H_{i,j-1}\right) - H_{i,j-1}
$$

and

$$
\begin{aligned}
H_{i,j-1} & = [y_{j-1} - d_{j-1} + H_{i,j-2}]^+ \\
& = \max\left(d_{j-1} - y_{j-1}, H_{i,j-2}\right) + y_{j-1} - d_{j-1}.
\end{aligned}
$$

The last equation in (10.5) is due to

$$
\begin{aligned}
& \max\left(d_{i,j-1} - y_{i,j-1}, d_{i,j-2} - y_{i,j-2}, \cdots, d_i - y_i, 0\right) \\
= \quad & M(i,j) - (y_{i,j} - d_{i,j}). \qquad \square
\end{aligned}
$$

We can now present the objective function that we want to maximize.

$$\sum_{i=1}^{N} \big\{ - c_i(y_i - x_i) + \mathsf{E}\big[\, p_i \min(D_i, y_i)$$

$$+ p_i \min\{(D_i - y_i)^+, H_{a^{-1}(i),i-1}\}$$
$$+ v_i(y_i - D_i - S_{i+1,a(i)})^+$$
$$- q_i(D_i - y_i - H_{a^{-1}(i),i-1})^+ \,\big] \big\}. \tag{10.8}$$

The first term (under the summation) is the acquisition cost (for bringing the on-hand inventory from $\mathbf{x}$ to $\mathbf{y}$), the second term is the revenue from supplying demand directly (i.e., using the same type of products), the third term is the revenue from supplying demand with substitution, the fourth term is the net salvage value (i.e., salvage value minus inventory holding cost; recall, $v_i := s_i - h_i$), and the fifth term is the penalty cost for shortage.

Notice that

$$\min\big[(D_i - y_i)^+, H_{a^{-1}(i),i-1}\big]$$
$$= (D_i - y_i)^+ - (D_i - y_i - H_{a^{-1}(i),i-1})^+$$

and

$$\min(D_i, y_i) + (D_i - y_i)^+ = D_i.$$

The expectation part in (10.8), the objective function, can be rewritten as

$$\mathsf{E}\big\{ p_i D_i + v_i[y_i - D_i - S_{i+1,a(i)}]^+ - r_i[D_i - y_i - H_{a^{-1}(i),i-1}]^+ \big\}.$$

(Recall that $r_i = p_i + q_i$.) Making use of the notation in Lemma 10.5, we can rewrite this as

$$\mathsf{E}\big\{ r_i y_i - q_i D_i + v_i[\tilde{M}(i, a(i)) - \tilde{M}(i+1, a(i))]$$
$$- r_i[\tilde{M}(a^{-1}(i), i) - \tilde{M}(a^{-1}(i), i-1)] \big\},$$

where $\tilde{M}$ denotes M with $\mathbf{d}$ replaced by the random demand $\mathbf{D}$, i.e.,

$$\tilde{M}(i, j) := \max\,(y_{i,j} - D_{i,j}, y_{i+1,j} - D_{i+1,j}, \cdots, y_j - D_j, 0).$$

Hence, the maximization problem can be presented as follows:

$$J(\mathbf{x}) = \max_{\mathbf{y} \geq \mathbf{x}} V(\mathbf{y}|\mathbf{x}) := \sum_{i=1}^{N} c_i x_i + \mathsf{E}[\tilde{G}(\mathbf{y})], \tag{10.9}$$

where

$$\tilde{G}(\mathbf{y}) := \sum_{i=1}^{N} [(r_i - c_i)y_i - q_i D_i] + \sum_{i=1}^{N} v_i[\tilde{M}(i, a(i)) - \tilde{M}(i+1, a(i))]$$

$$- \sum_{i=1}^{N} r_i[\tilde{M}(a^{-1}(i), i) - \tilde{M}(a^{-1}(i), i-1)]. \tag{10.10}$$

To simplify the notation, we will ignore $\mathbf{x}$ in $V(\mathbf{y}|\mathbf{x})$ and denote it as $V(\mathbf{y})$ in the following discussion. We must keep in mind, however, the dependence of V on the initial inventory $\mathbf{x}$, too.

10.4   Concavity and Submodularity

Here we want to show that $\mathsf{E}[\tilde{G}(\mathbf{y})]$ defined in (10.10) (and hence the objective function $V(\mathbf{y})$) is concave and submodular in $\mathbf{y}$. Note that the first term is linear in $\mathbf{y}$ and hence is trivially both concave and submodular, so it can be ignored. Furthermore, no generality is lost by replacing $\mathbf{D}$ with its realization $\mathbf{d}$. Hence, we define

$$G(\mathbf{y}) \quad := \quad \sum_{i=1}^{N} v_i[M(i, a(i)) - M(i+1, a(i))]$$

$$- \sum_{i=1}^{N} r_i[M(a^{-1}(i), i) - M(a^{-1}(i), i-1)] \qquad (10.11)$$

and will prove that $G(\mathbf{y})$ is concave and submodular in $\mathbf{y}$.

Lemma 10.6 $M(i, j)$ as defined in Lemma 10.5 is convex and supermodular in $(y_i, ..., y_j)$.

Proof. Recall

$$M(i, j) = \max(y_{i,j} - d_{i,j}, y_{i+1,j} - d_{i+1,j}, \cdots, y_j - d_j, 0). \qquad (10.12)$$

(If $i > j$, $M(i, j) \equiv 0$ by definition.) Hence, convexity is immediate, because max is an increasing and convex function, and $y_{k,j}$ $(k = i, ..., j)$ are linear functions of $(y_i, ..., y_j)$.

To prove supermodularity, let the increments $\Delta_k > 0$ and $\Delta_l > 0$ be added, first, to both components y_k and y_l, with $i \leq k < l \leq j$; next, to y_k only; and finally, to y_l only. Denote the resulting $M(i, j)$ as

$$\begin{aligned}
\phi_{k,l} \quad = \quad &\max(y_{i,j} - d_{i,j} + \Delta_k + \Delta_l, \cdots, y_{k,j} - d_{k,j} + \Delta_k + \Delta_l, \\
&y_{k+1,j} - d_{k+1,j} + \Delta_l, \cdots, y_{l,j} - d_{l,j} + \Delta_l, \\
&y_{l+1,j} - d_{l+1,j}, \cdots, y_j - d_j, 0),
\end{aligned}$$

$$\begin{aligned}
\phi_k \quad = \quad &\max(y_{i,j} - d_{i,j} + \Delta_k, \cdots, y_{k,j} - d_{k,j} + \Delta_k, \\
&y_{k+1,j} - d_{k+1,j}, \cdots, y_j - d_j, 0),
\end{aligned}$$

and

$$\begin{aligned}
\phi_l \quad = \quad &\max(y_{i,j} - d_{i,j} + \Delta_l, \cdots, y_{l,j} - d_{l,j} + \Delta_l, \\
&y_{l+1,j} - d_{l+1,j}, \cdots, y_j - d_j, 0).
\end{aligned}$$

Also, write $\phi_0 = M(i,j)$, i.e., without the addition of either increment. We want to prove
$$\phi_{k,l} - \phi_l \geq \phi_k - \phi_0.$$

If ϕ_k reaches its maximum at the term that does not involve Δ_k, then $\phi_k - \phi_0 = 0$, and the inequality obviously holds (because max is increasing) true. Otherwise, suppose
$$\phi_k = y_{k_1,j} - d_{k_1,j} + \Delta_k$$

for some $k_1 \leq k$. Then
$$\phi_{k,l} = y_{k_1,j} - d_{k_1,j} + \Delta_k + \Delta_l.$$

Hence,
$$\phi_{k,l} - \phi_k = \Delta_l \geq \phi_l - \phi_0,$$

where the inequality follows from
$$\phi_l \leq \max(y_{i,j} - d_{i,j} + \Delta_l, \cdots, y_j - d_j + \Delta_l, \Delta_l) = \phi_0 + \Delta_l. \qquad \square$$

We will write $G(\mathbf{y})$ as a linear combination of $M(i,j)$ with nonpositive coefficients; the desired concavity and submodularity then follow from Lemma 10.6.

Define two sets of indices as follows:
$$1 := i_1 < i_2 < \cdots < i_k \leq N, \quad 1 \leq j_1 < j_2 < \cdots < j_k := N, \qquad (10.13)$$

and $j_0 := 0$, $i_{k+1} := N + 1$, such that for all $\ell = 1, ..., k$,
$$a(n) = j_\ell, \ i_\ell \leq n \leq i_{\ell+1} - 1, \qquad (10.14)$$

i.e., the $\{i_\ell\}$ indices divide the N product types into subgroups so that each group consists of a contiguous subset of types, which shares a common $a(\cdot)$ value. For example, for the group $\{i_\ell, ..., i_{\ell+1} - 1\}$, the common $a(\cdot)$ value is j_ℓ [cf. (10.14)]. This has an important implication: this group will not (profitably) substitute for any type above j_ℓ. From this definition, it is easy to see that we also have
$$a^{-1}(n) = i_\ell, \ j_{\ell-1} < n \leq j_\ell. \qquad (10.15)$$

An example will better illustrate the idea behind this notation. Recall that, based on the optimal substitution policy in Proposition 10.2, $i \to j$ denotes that type i can profitably substitute for type j (i.e., $v_i \leq r_j$). Suppose $N = 6$ and the data are such that
$$1 \to 2, 3; \quad 2 \to 3, 4; \quad 3 \to 4; \quad 4 \to 5, 6; \quad 5 \to 6.$$

Then, following the notation in (10.13), we have $k = 3$, with

$$i_1 = 1, \quad i_2 = 2, \quad i_3 = 4; \quad \text{and} \quad j_1 = 3, \quad j_2 = 4, \quad j_3 = 6.$$

With these notations, we can now rewrite the first summation of (10.11) as follows:

$$\sum_{i=1}^{N} v_i[M(i, a(i)) - M(i+1, a(i))]$$
$$= \quad v_1 M(1, j_1) + (v_2 - v_1)M(2, j_1) + \cdots + (v_{i_2-1} - v_{i_2-2})M(i_2 - 1, j_1)$$
$$- v_{i_2-1}M(i_2, j_1) + v_{i_2}M(i_2, j_2) + (v_{i_2+1} - v_{i_2})M(i_2 + 1, j_2) + \cdots$$
$$- v_{i_3-1}M(i_3, j_2) + \cdots + v_{i_k}M(i_k, N) + (v_{i_k+1} - v_{i_k})M(i_k + 1, N)$$
$$+ \cdots + (v_{N-1} - v_{N-2})M(N - 1, N) + (v_N - v_{N-1})M(N, N).$$

Regrouping terms, we have

$$\sum_{i=1}^{N} v_i[M(i, a(i)) - M(i+1, a(i))]$$
$$= \quad v_1 M(1, j_1) + \sum_{\ell=2}^{k}[v_{i_\ell}M(i_\ell, j_\ell) - v_{i_\ell-1}M(i_\ell, j_{\ell-1})]$$
$$+ \sum_{\ell=1}^{k} \sum_{i=i_\ell+1}^{i_{\ell+1}-1} (v_i - v_{i-1})M(i, j_\ell). \tag{10.16}$$

Similarly, the second summation in (10.11) can be written as

$$- \sum_{i=1}^{N} r_i[M(a^{-1}(i), i) - M(a^{-1}(i), i - 1)]$$
$$= \quad -r_{j_1} M(1, j_1) + \sum_{\ell=2}^{k}[r_{j_{\ell-1}}M(i_\ell, j_{\ell-1}) - r_{j_\ell}M(i_\ell, j_\ell)]$$
$$+ \sum_{\ell=1}^{k} \sum_{j=j_{\ell-1}+1}^{j_\ell-1} (r_{j+1} - r_j)M(i_\ell, j). \tag{10.17}$$

Theorem 10.7 The objective function in (10.9), $V(\mathbf{y})$, is concave and submodular in $\mathbf{y}$.

Proof. As discussed earlier, it suffices to show that $G(\mathbf{y})$ of (10.11) is concave and submodular in $\mathbf{y}$. From (10.16), the coefficients of the Ms in the double summation are all nonpositive, due to Assumption 10.1 (ii); and the same is true for the coefficients of the Ms in the double summation in (10.17). Combine the other terms in (10.16) and (10.17) according to the arguments of the Ms; the resulting coefficients are all nonpositive as well. This is because, following Proposition 10.2, we know that for $\ell = 1, ..., k$, $a(i_\ell) = j_\ell$ [cf. (10.15)] implies $v_{i_\ell} \leq r_{j_\ell}$, and $i_\ell - 1 < i_\ell = a^{-1}(j_{\ell-1} + 1)$ [cf. (10.14)] implies $v_{i_\ell - 1} \geq r_{j_{\ell-1}+1}$. Because $G(\mathbf{y})$ is a linear combination of the Ms, with all the coefficients nonpositive, the desired concavity and submodularity follow from Lemma 10.6. $\square$

10.5   The Optimal Order Quantities

The optimal order quantities are obtained through solving the optimization problem in (10.9), $\max_{\mathbf{y} \geq \mathbf{x}} V(\mathbf{y})$. This is easily solved as a concave program, using, for example, a standard Lagrangian multiplier approach. We will study more about the properties of the optimal solution.

Let $\mathbf{z}^* := \arg\max V(\mathbf{y})$. That is, $\mathbf{z}^*$ is the solution to the maximization problem in (10.9), *without* the constraint $\mathbf{y} \geq \mathbf{x}$. Because $V(\mathbf{y})$ is concave, $\mathbf{z}^*$ is well defined, and we assume there is a solution procedure that obtains $\mathbf{z}^*$. We want to relate to $\mathbf{z}^*$ the optimal solution to (10.9), with the constraint.

Denote $A = \{i : z_i^* < x_i\}$ and let $\bar{A}$ be the complement of A. Allow $A = \emptyset$, the empty set. Write

$$\mathbf{z}_A^* := (z_i^*)_{i \in A} \quad \text{and} \quad \mathbf{z}_{\bar{A}}^* := (z_i^*)_{i \in \bar{A}}. \tag{10.18}$$

Similarly denote $\mathbf{x}_A$, $\mathbf{x}_{\bar{A}}$, $\mathbf{y}_A$, $\mathbf{y}_{\bar{A}}$, and so forth.

Proposition 10.8 Suppose $\mathbf{y}^*$ is an optimal solution to (10.9). If $\mathbf{y}_{\bar{A}}^* \geq \mathbf{z}_{\bar{A}}^*$, then both $\mathbf{y}^1 = (\mathbf{y}_A^*, \mathbf{z}_{\bar{A}}^*)$ and $\mathbf{y}^2 = (\mathbf{x}_A, \mathbf{y}_{\bar{A}}^*)$ are also optimal solutions to (10.9).

Proof.
  Because

$$\mathbf{x}_A \leq \mathbf{y}_A^*, \qquad \mathbf{z}_{\bar{A}}^* \leq \mathbf{y}_{\bar{A}}^*,$$

submodularity implies that

$$
\begin{aligned}
V(\mathbf{z}^*) + V(\mathbf{y}^*) &= V(\mathbf{z}_A^*, \mathbf{z}_{\bar{A}}^*) + V(\mathbf{y}_A^*, \mathbf{y}_{\bar{A}}^*) \\
&\leq V(\mathbf{z}_A^*, \mathbf{y}_{\bar{A}}^*) + V(\mathbf{y}_A^*, \mathbf{z}_{\bar{A}}^*) \\
&= V(\mathbf{z}_A^*, \mathbf{y}_{\bar{A}}^*) + V(\mathbf{y}^1).
\end{aligned}
$$

The maximality of the two terms on the left-hand side implies that the inequality must hold as an equality, and in particular $\mathbf{y}^1$ must also be an optimal solution.

  To argue for the optimality of $\mathbf{y}^2$, note that the optimality of $\mathbf{z}^*$ implies:

$$V(\mathbf{z}_A^* + \mathbf{y}_A^* - \mathbf{x}_A, \mathbf{z}_{\bar{A}}^*) - V(\mathbf{z}_A^*, \mathbf{z}_{\bar{A}}^*) \leq 0.$$

From the submodularity of $V(\cdot)$, we have

$$V(\mathbf{z}_A^* + \mathbf{y}_A^* - \mathbf{x}_A, \mathbf{y}_{\bar{A}}^*) - V(\mathbf{z}_A^*, \mathbf{y}_{\bar{A}}^*)$$
$$\leq\ V(\mathbf{z}_A^* + \mathbf{y}_A^* - \mathbf{x}_A, \mathbf{z}_{\bar{A}}^*) - V(\mathbf{z}_A^*, \mathbf{z}_{\bar{A}}^*)$$
$$\leq\ 0,$$

which, along with the concavity of $V(\cdot)$, yields

$$V(\mathbf{y}_A^*, \mathbf{y}_{\bar{A}}^*) - V(\mathbf{x}_A, \mathbf{y}_{\bar{A}}^*)$$
$$=\ V(\mathbf{x}_A + \mathbf{y}_A^* - \mathbf{x}_A, \mathbf{y}_{\bar{A}}^*) - V(\mathbf{x}_A, \mathbf{y}_{\bar{A}}^*)$$
$$\leq\ V(\mathbf{z}_A^* + \mathbf{y}_A^* - \mathbf{x}_A, \mathbf{y}_{\bar{A}}^*) - V(\mathbf{z}_A^*, \mathbf{y}_{\bar{A}}^*)$$
$$\leq\ 0.$$

Because $\mathbf{y}^*$ is optimal, $\mathbf{y}^2 = (\mathbf{x}_A, \mathbf{y}_{\bar{A}}^*)$ must also be optimal. $\square$

Proposition 10.9 If $A = \emptyset$, then $\mathbf{z}^*$ is the optimal solution to (10.9). If $\bar{A} = \emptyset$, then $\mathbf{y}^* = \mathbf{x}$ is an optimal solutions to (10.9). Otherwise, there exists an optimal solution to (10.9), $\mathbf{y}^*$, that has the following properties:

(i) there exists at least one $i \in A$, such that $y_i^* = x_i$; and

(ii) there exists at least one $j \in \bar{A}$, such that $y_j^* \leq z_j$.

Proof. $A = \emptyset$ means $\mathbf{z}^*$ satisfies the constraint $\mathbf{z}^* \geq \mathbf{x}$ and hence must be an optimal solution to (10.9). On the other hand, when $\bar{A} = \emptyset$, from the proof of Proposition 10.8 it is clear that $\mathbf{y}^2 = \mathbf{x}$ is optimal. (In this case, $\mathbf{y}^1$ in the proposition becomes trivial: $\mathbf{y}^1 = \mathbf{y}^*$.)
    To prove (i), let

$$\lambda := \max_{j \in A} (x_j - z_j^*)/(y_j^* - z_j^*) = (x_i - z_i^*)/(y_i^* - z_i^*),$$

i.e., suppose the maximum is attained at some $i \in A$. Note that above λ exists and $0 \leq \lambda \leq 1$. Consider the convex combination:

$$\tilde{\mathbf{y}} := \lambda \mathbf{y}^* + (1 - \lambda)\mathbf{z}^*.$$

Note that $\tilde{\mathbf{y}}_A \geq \mathbf{x}_A$ and, in particular, $\tilde{y}_i = x_i$. The concavity of V implies:

$$\lambda V(\mathbf{y}^*) + (1 - \lambda)V(\mathbf{z}^*) \leq V(\tilde{\mathbf{y}}). \tag{10.19}$$

On the other hand, the maximality of $V(\mathbf{z}^*)$ on the left-hand side implies that $V(\mathbf{y}^*) \leq V(\tilde{\mathbf{y}})$. Hence, $\tilde{\mathbf{y}}$ must also be an optimal solution, which satisfies $y_i = x_i$.
    Part (ii) follows immediately from Proposition 10.8. If $y_j^* > z_j^*$ for all $j \in \bar{A}$, then $\mathbf{y}^1$ in Proposition 10.8 is optimal, in particular, $\mathbf{y}_{\bar{A}}^1 = \mathbf{z}_{\bar{A}}^*$. $\square$
    Notice that, when $0 < \lambda < 1$, (10.19) implies

$$V(\mathbf{y}^*) = V(\mathbf{z}^*) = V(\mathbf{y}^2).$$

We have the following.

Corollary 10.10 If $A = \emptyset$, or there exists an optimal solution $\mathbf{y}^*$ such that $y_i^* \neq x_i$ for all $i \in A$, then $V(\mathbf{y}^*) = V(\mathbf{z}^*)$.

Applying the properties of Proposition 10.9 to the case of two products, we have the following.

Proposition 10.11 In the case of $N = 2$ product types, the optimal order quantity $\mathbf{y}^* = (y_1^*, y_2^*)$ relates to $\mathbf{z}^* = (z_1^*, z_2^*)$ as follows:

(i) if $\mathbf{z}^* \geq \mathbf{x}$, then $\mathbf{y}^* = \mathbf{z}^*$. If $\mathbf{z}^* << \mathbf{x}$, then $\mathbf{y}^* = \mathbf{x}$. (This holds even when $N > 2$.)

(ii) otherwise, if $z_1^* < x_1$ and $z_2^* \geq x_2$, then $y_1^* = x_1$ and $x_2 \leq y_2^* \leq z_2^*$; if $z_1^* \geq x_1$ and $z_2^* < x_2$, then $x_1 \leq y_1^* \leq z_1^*$ and $y_2^* = x_2$.

One might ask whether it is true that if $\mathbf{z}_A^* << \mathbf{x}_A$, the optimal solution $\mathbf{y}^*$ can be obtained by setting $\mathbf{y}_A^* = \mathbf{x}_A$ and then optimizing the $\bar{A}$ part. This clearly holds in $N = 2$, as is evident from Proposition 10.11, but unfortunately it does not hold in general. Consider an example of $N = 3$. Let $c_1 = 1.6$, $c_2 = 1.0$, $c_3 = .75$, $p_1 = 2.0$, $p_2 = 1.6$, $p_3 = 0.9$, $h_1 = h_2 = h_3 = 0.1$, $s_1 = 1.2$, $s_2 = 0.8$, $s_3 = 0.2$, $q_1 = 1.0$, $q_2 = 0.8$, and $q_3 = 0.3$. Demands are all uniformly distributed in $[10,50]$. Here $\mathbf{z}^* = (40, 42, 19)$ (rounded to integer values), while for $\mathbf{x} = (75, 0, 0)$, the optimal solution is $\mathbf{y}^* = (75, 28, 22)$. This means that for another vector of initial inventory, $\mathbf{x}' = (75, 0, 20)$, say, the optimal solution is still $\mathbf{y}^* = (75, 28, 22)$. In other words, keeping $x_3 = 20 > z_3 = 19$ is *not* good enough (not optimal).

10.6   Upper and Lower Bounds

Lemma 10.12 $M(i, j)$ as defined in (10.6) has the followings properties:

(i) $M(1, j) - M(2, j)$ is increasing in y_i, $i = 2, ..., j$;

(ii) $M(i, N) - M(i, N - 1)$ is decreasing in y_j, $j = i + 1, ..., N - 1$.

Proof. (i) Suppose y_i, for some i between 2 and j, is increased by an amount $\delta > 0$, and denote the resulting M by M^δ. We want to show

$$M^\delta(1, j) - M^\delta(2, j) \geq M(1, j) - M(2, j). \tag{10.20}$$

Suppose the right-hand side is zero. Then the inequality obviously holds, because from (10.12), we have

$$M^\delta(1, j) = \max\{y_{1,j} - d_{1,j} + \delta, M^\delta(2, j)\}.$$

On the other hand, if the right-hand side of (10.20) is positive, then it must be that the maximum of $M(1, j)$ is reached at the first term, denoted

$M(1, j) = M_1$, while the maximum of $M(2, j)$ is reached at an ℓth term, denoted $M(2, j) = M_\ell$, with $2 \leq \ell \leq j$. Adding δ to y_i makes $M^\delta(1, j) = M_1 + \delta$, because M_1 involves y_i. On the other hand, this clearly cannot increase $M(2, j)$ by more than δ, i.e., $M^\delta(2, j) \leq M_\ell + \delta$. Hence, the desired inequality.

(ii) Similar to (i), suppose now that y_j, for some j between i and $N - 1$, is increased by an amount $\delta > 0$, and denote the resulting M by M^δ. We want to show

$$M^\delta(i, N) - M^\delta(i, N - 1) \leq M(i, N) - M(i, N - 1). \tag{10.21}$$

Suppose $M(i, N - 1) = y_{\ell, N-1} - d_{\ell, N-1}$, where ℓ is some index between i and N (when $\ell = N$, $y_{\ell, N-1} = d_{\ell, N-1} = 0$). Denote $M_\ell := M(i, N - 1)$. Then the right-hand side of (10.21) can be expressed as

$$[M_\ell + y_N - d_N]^+ - M_\ell = \max(y_N - d_N, -M_\ell).$$

(Notice that every term in $M(i, N)$ under the max, with the exception of the zero term, is the corresponding term in $M(i, N - 1)$ plus $y_N - d_N$.) After δ is added to y_j, suppose $M^\delta(i, N - 1) = M_k^\delta$, i.e., the maximum is reached at the term subscripted k, where k is between i and $N - 1$ (k could be equal to ℓ). Then the left-hand side of (10.21) can be expressed as

$$[M_k^\delta + y_N - d_N]^+ - M_k^\delta = \max(y_N - d_N, -M_k^\delta).$$

Clearly, $M_k^\delta \geq M_\ell$. Hence, the left-hand side of (10.21) is indeed dominated by the right-hand side, i.e., the inequality holds. $\square$

Denote $G'(\mathbf{y}_1)$ as the $G(\mathbf{y})$ function [cf. (10.11)] corresponding to $N - 1$ types of products, with type 1 removed, in particular, $\mathbf{y}_1 = (y_2, ..., y_N)$. Similarly, denote $G'(\mathbf{y}_N)$, with type N products removed.

Lemma 10.13 (i) $G(\mathbf{y}) - G'(\mathbf{y}_1)$ is decreasing in $\mathbf{y}_1$.
(ii) $G(\mathbf{y}) - G'(\mathbf{y}_N)$ is increasing in $\mathbf{y}_N$.

Proof. (i) From (10.16) and (10.17), we have

$$\begin{aligned} G(\mathbf{y}) &- G'(\mathbf{y}_1) \\ = \ & (v_1 - r_{j_1})[M(1, j_1) - M(2, j_1)] \\ & + \sum_{j=1}^{j_1-1} (r_{j+1} - r_j)[M(1, j) - M(2, j)]. \end{aligned}$$

Because $v_1 \leq r_{j_1}$ (recall $i_1 = 1$) and $r_{j+1} \leq r_j$, the desired decreasing property follows from Lemma 10.12 (i).

(ii) Similarly, from (10.16) and (10.17), we have

$$G(\mathbf{y}) - G'(\mathbf{y}_N)$$
$$= (v_{i_k} - r_N)[M(i_k, N) - M(i_k, N-1)]$$
$$+ \sum_{i=i_k+1}^{N} (v_i - v_{i-1})[M(i, N) - M(i, N-1)].$$

Because $v_{i_k} \leq r_N$ (recall $j_k = N$) and $v_i \leq v_{i-1}$, the desired increasing property follows from Lemma 10.12 (ii). $\square$

An immediate consequence of Lemma 10.13 is the following.

Proposition 10.14 Let $y_{(i)}$, for $i = 1$ and N, be the solution to the single-product problem, with $i = 1$ or N being the product type. Then $y_{(1)} \leq y_1^*$ and $y_{(N)} \geq y_N^*$, where y_1^* and y_N^* are the optimal order quantities in the N-product problem.

10.7   Notes

Pentico [69] studies a substitution problem that is similar to our model. The partial substitution policy there is 'segment substitution', which is predetermined. McGillivray and Silver [59] investigated the effects of substitutability on stocking control rules and inventory/storage costs for the case where all items have the same unit variable cost and shortage penalty. The stocking control rule is (R, S_i), i.e., every R period the stock of each product is raised to the order-up-to level S_i, $i = 1, \cdots, N$. Parlar and Goyal [66] study a two-product single-period substitution problem. Pasternack and Drezner [68] consider the same model as the one in [66], but with different revenue levels for the two products. Several cases are compared, including two-way substitution (each product can be used as a substitute for the other), one-way substitution (one product can be used as a substitute for the other but not vice versa), and no substitution (neither product can be used to substitute for the other). Bassok et al. [5] extend the model of Pasternack and Drezner to the general multiproduct case with a predetermined serial full-substitution mechanism. Under several assumptions, the objective function is shown to be both concave and submodular. Other related models include Hsu and Bassok [44], Klein, Luss, and Rothblum [50], and Robinson [72], among others.

In most earlier studies, the rule of substitution is prespecified. However, the substitution rule itself is an important aspect of the problem. The model studied in this chapter, which draws materials from [18], is an extension of Bassok et al. [5]. We allow partial substitution, and the substitution rule is not predetermined but rather is optimized, taking into account all the relevant cost parameters. With this optimal substitution rule, the objective function is shown to be both concave and submodular.

Extending our model to multiple periods is possible, although establishing concavity and submodularity becomes substantially more challenging. In [19], we established these results for the case of two product types. Earlier works on this topic include such classics as Veinott [99], and Ignall and Veinott [46], where the underlying mathematical structure is elegantly outlined.

References

[1] ALBIN, S.L., AND FRIEDMAN, D.J., The Impact of Clustered Defect Distributions in IC Fabrication, *Mgmt. Sci.*, **35** (1989), 1066-78.

[2] ANUPINDI, R., AND AKELLA, R., Diversification under Supply Uncertainty, *Mgmt. Sci.*, **39** (1993), 944-63.

[3] BALLOU, D., AND PAZER, H., The Impact of Inspector Facility on the Inspection Policy in Serial Production Systems, *Mgmt. Sci.*, **28** (1982), 387-99.

[4] BANKS, J., *Principles of Quality Control*, Wiley, New York, 1989.

[5] BASSOK, Y., ANUPINDI, R., AND AKELLA, R., Single-Period Multi-product Inventory Models with Substitution, *Operations Research*, **47** (1999), 632-42.

[6] BERTSEKAS, D.P., *Dynamic Programming and Optimal Control*, Athena Scientific, Belmont, MA, 1995.

[7] BEUTLER, F.J., AND ROSS, K.W., Optimal Policies for Controlled Markov Chains with a Constraint, *J. of Math. Ana. & Appl.*, **112** (1985), 236-52.

[8] BLACKWELL, D., Discounted Dynamic Programming, *Ann. of Math. Statist.*, **36** (1965), 226-35.

[9] BLISCHKE, W.R., Mathematical Models for Analysis of Warranty Policies, *Math. Comput. Modeling*, **13** (1990), 1-16.

[10] BRITNEY, R.R., Optimal Screening Plans for Non-Serial Production Systems, *Mgmt. Sci.*, **18** (1972), 550-9 .

[11] BUZACOTT, J.A., AND SHANTHIKUMAR, J.G., *Stochastic Models of Manufacturing Systems*, Prentice-Hall, Englewood Cliffs, NJ, 1993.

[12] BUZACOTT, J.A., AND SHANTHIKUMAR, J.G., Design of Manufacturing Systems Using Queueing Models, *Queueing Systems*, **12** (1992), 890-905.

[13] CASSANDRAS, C.G., Optimal Policies for the "Yield Learning" Problem in Manufacturing Systems, *IEEE Trans. on Automatic Control*, **AC-41** (1996), 1210-13.

[14] CHANG, C.S., SHANTHIKUMAR, J.G., AND YAO, D.D., Stochastic Convexity and Stochastic Majorization, in *Stochastic Modeling and Analysis of Manufacturing Systems* (Chapter 5), D.D. Yao (ed.) Springer-Verlag, New York, 1994.

[15] CHANG, C.S., AND YAO, D.D., Rearrangement, Majorization, and Stochastic Scheduling. *Mathematics of Operations Research*, **18** (1993), 658-84.

[16] CHEN, H., HARRISON, M.J., ACKERE A.V., AND WEIN L.M., Empirical Evaluation of a Queueing Network Model for Semiconductor Wafer Fabrication, *Operations Research*, **36** (1988), 202-15.

[17] CHEN, H., YANG, P., AND YAO, D.D., Control and Scheduling in a Two-Station Network: Optimal Policies and Heuristics, *Queueing Systems*, **18** (1994), 301-32.

[18] CHEN, J., *Substitution and Inspection Models in Production-Inventory Systems*, Ph.D. Dissertation, IEOR Dept., Columbia University, New York, 1997.

[19] CHEN, J., YAO, D.D., AND ZHENG, S., A Multi-product Multi-period Inventory Model with Substitution, Working paper (2001).

[20] CHEN, J., YAO, D.D., AND ZHENG, S., Optimal Replenishment and Rework with Multiple Unreliable Supply Sources, *Operations Research*, **49** (2001), 430-43.

[21] CHEN, J., YAO, D.D., AND ZHENG, S., Quality Control for Products Supplied with Warranty, *Operations Research*, **46** (1998), 107-15.

[22] CHENG, D.W., AND YAO, D.D., Tandem Queues with General Blocking: A Unified Model and Stochastic Comparisons, *Discrete Event Dynamic Systems: Theory and Applications*, **2** (1993), 207-34.

[23] CHITTAYIL, K., KUMAR, R.T., AND COHEN, P.H., Acoustic Emission Sensing for Tool Wear Monitoring and Process Control in Metal Cutting, in *Handbook of Design, Manufacturing and Automation* (Chapter 33), R.C. Dorf and A. Kusiak (eds.), Wiley, New York, 1994.

[24] CHOW, Y.S., ROBBINS, H., AND SIEGMUND, D., *Great Expectations, The Theory of Optimal Stopping,* Houghton Mifflin, New York, 1972.

[25] COLEMAN, D.E., Generalized Control Charting, in *Statistical Process Control in Automated Manufacturing,* J.B. Keats and N.F. Hubele (eds.), Marcel Dekker, New York, 1989.

[26] CONNORS, D., FEIGIN, G., AND YAO, D.D., A Queueing Network Model for Semiconductor Manufacturing, *IEEE Trans. on Semiconductor Manufacturing,* **9** (1996), 412-27.

[27] CONNORS, D., FEIGIN, G., AND YAO, D.D., Scheduling Semiconductor Lines Using a Fluid Network Model, *IEEE Trans. on Robotics and Automation,* **10** (1994), 88-98.

[28] CROWDER, S.V., An Application of Adaptive Kalman Filtering to Statistical Process Control, in *Statistical Process Control in Automated Manufacturing,* J.B. Keats and N.F. Hubele (eds.), Marcel Dekker, New York, 1989.

[29] DEMING, W.E., Foreword to *Statistical Method from the Viewpoint of Quality Control,* W.A. Shewhart, Dover Publications, New York, 1986.

[30] DEMING, W.E., *Out of the Crisis,* MIT Press, Cambridge, MA, 1986.

[31] DERMAN, C., *Finite State Markovian Decision Processes,* Academic Press, New York, 1970.

[32] DJAMALUDIN, I., MURTHY, D.N.P., AND WILSON, R.J., Quality Control through Lot Sizing for Items Sold with Warranty, *International J. of Production Economics,* **33** (1994), 97-107.

[33] EPPEN, G.D., AND HURST, E.G., Optimal Location of Inspection Stations in a Multistage Production Process, *Mgmt. Sci.,* **20** (1974), 1194-1200.

[34] FEDERGRUEN A., AND TIJMS, H.C., The Optimality Equation in Average Cost Denumerable State Semi-Markov Decision Problems, Recurrency Conditions and Algorithms, *J. Appl. Prob.,* **15** (1978), 356-73.

[35] FEINBERG, E.A., Constrained Semi-Markov Decision Processes with Average Rewards, *ZOR–Mathematical Methods of Operations Research,* **39** (1994), 257-88.

[36] GERCHAK, Y., AND PARLAR, M., Yield Randomness/Cost Tradeoffs and Diversification in the EOQ Model, *Naval Research Logistics,* **37** (1990), 341-54.

[37] GIRSHICK, M.A., AND RUBIN H., A Bayes Approach to a Quality Control Model, *Ann. of Math. Statist.,* **23** (1952), 114-25.

[38] GLASSERMAN, P., AND WANG, Y., Leadtime-Inventory Tradeoffs in Assemble-to-Order Systems, *Operations Research,* **46** (1998), 858-71.

[39] GLASSERMAN, P., AND YAO, D.D., Structured Buffer Allocation Problems, *Discrete Event Dynamic Systems: Theory and Applications,* **6** (1996), 9-29.

[40] GLASSERMAN, P., AND YAO, D.D., *Monotone Structure in Discrete-Event Systems,* Wiley, New York, 1994.

[41] GLASSERMAN, P., AND YAO, D.D., Monotone Optimal Control of Permutable GSMPs, *Mathematics of Operations Research,* **19** (1994), 449-476.

[42] GLASSERMAN, P. AND YAO, D.D., A GSMP Framework for the Analysis of Production Lines, in *Stochastic Modeling and Analysis of Manufacturing Systems* (Chapter 4), D.D. Yao (ed.) Springer-Verlag, New York, 1994.

[43] GUNTER, S.I., AND SWANSON, L.A., Inspector Location in Convergent Production Lines, *Int. J. Prod. Res.* **23** (1985), 1153-69.

[44] HSU, A., AND BASSOK, Y., Random Yield and Random Demand in a Production System with Downward Substitution, *Operations Research,* **47** (1999), 277-90.

[45] HUBELE, N.F., A Multivariate and Stochastic Framework for Statistical Process Control, in *Statistical Process Control in Automated Manufacturing,* J.B. Keats and N.F. Hubele (eds.), Marcel Dekker, New York, 1989.

[46] IGNALL, E., AND VEINOTT, A.F., Optimality of Myopic Inventory Policies for Several Substitute Products, *Mgmt. Sci.,* **15** (1969), 284-304.

[47] KALLENBERG, L., *Linear Programming and Finite Markovian Control Problems,* Math Centre Tracts 148, Mathematisch Centrum, Amsterdam, 1983.

[48] KAMAE, T., KRENGEL, U., AND O'BRIEN, G.L., Stochastic Inequalities on Partially Ordered Spaces, *Ann. Probability,* **5** (1977), 899-912.

[49] KEILSON, J., AND SUMITA, U., Uniform Stochastic Ordering and Related Inequalities, *Canad. J. Statist.,* **10** (1982), 181-98.

[50] KLEIN, R.S., LUSS, H., AND ROTHBLUM, U.G., Minimax Resource Allocation Problems with Resource Substitutions Represented by Graphs, *Operations Research,* **41** (1993), 959-71.

[51] KUMAR, P.R., Re-Entrant Lines, *Queueing Systems, Theory and Applications,* **13** (1993), 87-110.

[52] LEE, H.L., Lot Sizing to Reduce Capacity Utilization in a Production Process with Defective Items, Process Corrections and Rework, *Mgmt. Sci.,* **38** (1992), 1314-28.

[53] LEE, H.L. AND ROSENBLATT, M.J., Optimal Inspection and Ordering Policies for Products with Imperfect Quality, *IIE Trans.,* **17** (1985), 284-9.

[54] LEE, H.L., AND YANO, C.A., Production Control in Multistage Systems with Variable Yield Loss, *Operations Research,* **36** (1988), 269-78.

[55] LINDSAY, G.F., AND BISHOP, A., Allocation of Screening Inspection Effort: A Dynamic Programming Approach, *Mgmt. Sci.,* **10** (1964), 342-52.

[56] LIE, C.H., AND CHUN, Y.H., Optimum Single-Sample Inspection Plans for Products Sold under Free and Rebate Warranty, *IEEE Trans. on Reliability,* **R-36** (1987), 634-47.

[57] LONGTIN, M.D., WEIN, L.M., AND WELSCH, R.E., Sequential Screening in Semiconductor Manufacturing, II: Exploiting Spatial Dependence, *Operations Research,* **44** (1996), 196-205.

[58] MAMER, J.W., Discounted and Per Unit Costs of Products Warranty, *Mgmt. Sci.,* **33** (1987), 916-30.

[59] MCGILLIVRAY, A.R., AND SILVER, V., Some Concepts for Inventory Control under Substitutable Demands, *INFOR,* **16** (1978), 47-63.

[60] MITRA, A., *Fundamentals of Quality Control and Improvement,* Macmillan, New York, 1993.

[61] MONTGOMERY, D., *Introduction to Statistical Quality Control,* Wiley, New York, 1991.

[62] MURTHY, D.N.P., AND NGUYEN, D.G., Optimal Development Testing Policies for Products Sold with Warranty, *Reliability Engineering,* **19** (1987), 113-23.

[63] MURTHY, D.N.P., WILSON, R.J., AND DJAMALUDIN, I., Product Warranty and Quality Control, *Quality and Reliability Engineering*, **9** (1993), 431-43.

[64] NGUYEN, D.G., AND MURTHY, D.N.P., An Optimal Policy for Servicing Warranty, *J. Opl. Res. Soc.*, **37** (1986), 1081-88.

[65] OU, J., AND WEIN, L.M., Sequential Screening in Semiconductor Manufacturing, I: Exploiting Lot-to-Lot Variability, *Operations Research*, **44** (1996), 173-95.

[66] PARLAR, M., AND GOYAL, S.K., Optimal Ordering Decisions for Two Substitutable Products with Stochastic Demands, *OPSEARCH*, **21** (1984), 1-15.

[67] PARLAR, M., AND WANG, D., Diversification under Yield Randomness in Inventory Models, *European J. of Opnl. Res.*, **66** (1993), 52-64.

[68] PASTERNACK, B., AND DREZNER, Z., Optimal Inventory Policies for Substitutable Commodities with Stochastic Demand, *Naval Research Logistics*, **38** (1990), 221-40.

[69] PENTICO, D.W., The Assortment Problem with Probabilistic Demands, *Mgmt. Sci.*, **21** (1974), 286-90.

[70] PETER, M.H., SCHNEIDER, H., AND TANG, K., Joint Determination of Optimal Inventory and Quality Control Policy, *Mgmt. Sci.*, **34** (1988), 991-1004.

[71] PUTERMAN, M.L., *Markov Decision Processes*, Wiley, New York, 1994.

[72] ROBINSON, L., Optimal and Approximate Policies in Multiperiod Multilocation Inventory Models with Transhipments, *Operations Research*, **38** (1990), 278-95.

[73] ROSS, S.M., *Stochastic Processes*, 2nd ed., Wiley, New York, 1996.

[74] ROSS, S.M., *Introduction to Stochastic Dynamic Programming*, Academic Press, New York, 1983.

[75] ROSS, S.M., Quality Control under Markovian Deterioration, *Mgmt. Sci.*, **17** (1971), 587-96.

[76] ROSS, S.M., *Applied Probability Models with Optimization Applications*, Holden-Day, 1970.

[77] SCARF, H., The Optimality of (S, s) Policies in the Dynamic Inventory Problem, in *Mathematical Methods in the Social Sciences*, 196-202, K.J. Arrow, S. Karlin and P. Suppes (eds.), Stanford University Press, Stanford, CA, 1960.

[78] SCHWEITZER, P.J., Iterative Solution of the Function Equations for Undiscounted Markov Renewal Programming, *J. Math. Anal. Appl.*, **34** (1971), 495-501.

[79] SERFOZO, R.F., An Equivalence Between Discrete and Continuous Time Markov Decision Processes, *Operations Research*, **27** (1979), 616-20.

[80] SHAKED, M., AND SHANTHIKUMAR, J.G., Stochastic Convexity and Its Applications, *Adv. Appl. Prob.*, **20** (1988), 427-46.

[81] SHAKED, M., AND SHANTHIKUMAR, J.G., *Stochastic Orders and Their Applications*, Academic Press, San Diego, CA, 1994.

[82] SHANTHIKUMAR, J.G., AND YAO, D.D., Spatial-Temporal Convexity of Stochastic Processes and Applications, *Probability in the Engineering and Informational Sciences*, **6** (1992) 1-16.

[83] SHANTHIKUMAR, J.G., AND YAO, D.D., Bivariate Characterization of Some Stochastic Order Relations, *Adv. Appl. Prob.*, **23** (1991), 642-59.

[84] SHANTHIKUMAR, J.G., AND YAO, D.D., Strong Stochastic Convexity: Closure Properties and Applications, *J. of Applied Probability*, **28** (1991), 131-45.

[85] SHANTHIKUMAR, J.G., AND YAO, D.D., The Preservation of Likelihood Ratio Ordering under Convolution, *Stoch. Proc. Appl.*, **23** (1986), 259-67.

[86] SHEWHART, W.A., *Statistical Method from the Viewpoint of Quality Control*, Dover Publications, New York, 1986.

[87] SINGPURWALLA, N.D., AND WILSON, S., The Warranty Problem: Its Statistical and Game Theoretic Aspects, *SIAM Review*, **35**(1) (March, 1993), 17-42.

[88] SONG, J.S., On the Order Fill Rate in a Multi-item, Base-Stock Inventory System, *Operations Research*, **46** (1998), 831-45.

[89] SONG, J.S., XU, S., AND LIU, B., Order Fulfillment Performance Measures in an Assembly-to-Order System with Stochastic Leadtimes, *Operations Research*, **47** (1999), 131-49.

[90] STOYAN, D., *Comparison Methods for Queues and Other Stochastic Models*, Wiley, New York, 1983.

[91] TAGUCHI, G., ELSAYED, A., AND HSIANG, T., *Quality Engineering in Production Systems*, McGraw-Hill, New York, 1989.

[92] TAGUCHI, G., AND WU, Y., *Introduction to Off-Line Quality Control,* Central Japan Quality Control Association, Nagoya, Japan, 1979.

[93] TAKATA, S., OGAWA, M., BERTOK, P., OOTSUKA, P., MATUSHIMA, K., AND SATA, T., Real-Time Monitoring System of Tool Breakage Using Kalman Filtering, *Robotics and Computer Integrated Manufacturing,* **2** (1985) 33-40.

[94] TAPIERO, C.S., AND LEE, H.L., Quality Control and Product Servicing: A Decision Framework, *Eur. J. Oper. Res.,* **39** (1989), 261-73.

[95] TAYLOR H., Markovian Sequential Replacement Processes, *Ann. of Math. Statist.,* **36** (1965), 1677-94.

[96] THOMAS, M.U., Optimum Warranty Policies for Nonrepairable Items, *IEEE Trans. on Reliability,* **R-32** (1983), 282-8.

[97] THOMPSON, J.R., AND KORONACKI, J., *Statistical Process Control for Quality Improvement,* Chapman & Hall, New York, 1993.

[98] TOPKIS, D.M., Minimizing a Submodular Function on a Lattice, *Operations Research,* **26** (1978), 305-21.

[99] VEINOTT, A.F., Optimal Policy for a Multi-product, Dynamic, Nonstationary Inventory Problem, *Mgmt. Sci.,* **12** (1965), 206-22.

[100] WALD, A., *Sequential Analysis,* Dover Publications, Inc., 1973.

[101] WALKER, D.M.H., *Yield Simulation for Integrated Circuits,* Kluwer Academic Publishers, Norwell, MA, 1987.

[102] WHITE, L., Shortest Route Models for Allocation of Inspection Effort on a Production Line, *Mgmt. Sci.,* **15** (1969), 249-59.

[103] WHITE, L., The Analysis of a Simple Calss of Multistage Inspection Plans, *Mgmt. Sci.,* **9** (1966), 685-93.

[104] WOLFF, R.W., *Stochastic Modeling and the Theory of Queues,* Prentice-Hall, Englewood Cliffs, NJ, 1989.

[105] YAO, D.D., Optimal Run Quantities for an Assembly System with Random Yields, *IIE Trans.,* **20**(4) (1988), 399-403.

[106] YAO, D.D., AND ZHENG S. Sequential Inspection Under Capacity Constraints, *Operations Research,* **47** (1999), 410-21.

[107] YAO, D.D., AND ZHENG S. Coordinated Quality Control in a Two-Stage System, *IEEE Trans. on Automatic Control,* **44** (1999), 1166-79.

[108] YAO, D.D. AND ZHENG, S., Markov Decision Programming for Process Control in Batch Manufacturing, *Probability in the Engineering and Informational Sciences*, **12** (1998), 351-372.

[109] YAO, D.D., AND ZHENG, S., Coordinated Production and Inspection in a Tandem System, *Queueing Systems*, **24** (1996), 59-82.

[110] ZHENG, S., Dynamic Quality Control in an Assembling Line, *IIE Trans.*, **32** (2000), 797-806.

Index